Next-Generation High-Speed Satellite Interconnect

Pietro Nannipieri • Gianmarco Dinelli
Luca Dello Sterpaio • Antonino Marino
Luca Fanucci

Next-Generation High-Speed Satellite Interconnect

Disclosing the SpaceFibre Protocol – A System Perspective

Pietro Nannipieri
Department of Information Engineering
University of Pisa
Pisa, Italy

Gianmarco Dinelli
IngeniArs S.r.l.
Pisa, Italy

Luca Dello Sterpaio
IngeniArs S.r.l.
Pisa, Italy

Antonino Marino
IngeniArs S.r.l.
Pisa, Italy

Luca Fanucci
Department of Information Engineering
University of Pisa
Pisa, Italy

ISBN 978-3-030-77046-4 ISBN 978-3-030-77044-0 (eBook)
https://doi.org/10.1007/978-3-030-77044-0

This Springer imprint is published by the registered company Springer Nature Switzerland AG
The registered company address is: Gewerbestrasse 11, 6330 Cham, Switzerland

Foreword

Today, many spacecraft incorporate SpaceWire for the communication between on-board equipment such as scientific payloads, mass memories, and on-board computers. The protocol was standardized in an open European Cooperation for Space Standardization (ECSS) standard in 2003, with support from all major space agencies (ESA, NASA, JAXA, and Roscosmos).

Already around that time the idea of a successor protocol was born with the overall objective of increasing the capabilities of SpaceWire by an order of magnitude due to higher data rates, improved Quality-of-Service (QoS), better Fault Detection, Isolation and Recovery (FDIR) capabilities, and the possibility of utilizing also optical fibres as communication links. This protocol, dubbed SpaceFibre, is the result of more than 15 years of research and development leading to an open ECSS standard finally released in 2019. The rather long development time is not untypical for the space industry, which often conservatively adapts new technologies due to the stringent reliability requirements of space missions. Because of this long development time, SpaceFibre represents a very robust and well-tested protocol today, with many interesting features particularly suited for space-borne equipment.

The ECSS standard and the first reference designs were developed by the University of Dundee and STAR-Dundee UK for ESA, with many valuable contributions from international space agencies, large system integrators, and research institutes. One of the European academic institutions particularly worth mentioning in this context is the University of Pisa. Several researchers contributed to the ECSS standard working group and have been researching SpaceFibre-related topics for several years now, with many interesting outcomes such as the development of end-point and routing switch IP cores, an Electrical Ground Support Equipment (EGSE) system, and a simulator for SpaceFibre networks.

This book, which is to our knowledge the first of its kind, gives an excellent overview of the SpaceFibre technology and is therefore well suited for anyone who wants to quickly understand the workings of the protocol. In addition, it also delves into implementation details of the different building blocks developed by the research group in Pisa over the previous years, revealing a wealth of practical information that should prove useful and inspirational to anyone working and researching on this topic.

ESA On-Board Computer & Data Handling Section Staff Member, European Space Agency, European Space Research and Technology Centre (ESTEC), Noordwijk, The Netherlands

Felix Siegle

Head of ESA Electrical Engineering Department, European Space Agency, European Space Research and Technology Centre (ESTEC), Noordwijk, The Netherlands

Riccardo De Gaudenzi

Preface

Earth observation services are vital to our society: all major and minor space agencies are massively investing resources to have better performance in monitoring Earth with satellites. Moreover, in the last few years, a wide range of private companies is also entering the business. Thanks to higher resolution instruments, Earth observation satellites' capabilities are consistently enhanced. However, to fully support the increasing demand for high-performance and high-bandwidth instruments onboard, an unprecedented requirement for faster and faster communication between instruments and mass memory is arising. For this reason, ESA recently proposed the SpaceFibre protocol (several Gbps bandwidth) for satellite onboard communication, as the successor of SpaceWire (hundreds of Mbps bandwidth). This book aims at giving a complete vision of the SpaceFibre protocol, together with an analysis of all the necessary hardware and software components to integrate this technology onboard a satellite. This book is addressed to all players involved in onboard satellite communication, from researcher to industry. The text provides a system perspective for the end user willing to adopt this technology for a future space mission, guiding potential system adopter in the understanding of the protocol, analysing strengths, weaknesses and performances. Also, practical design examples and prototype performance measurements in reference scenarios are included. The goal of the book is to introduce all space community members, both from academia and industry to this novel protocol. Indeed, SpaceFibre is expected to follow the success of its predecessor SpaceWire protocol (Mbps), which has been adopted by all significant space agencies (e.g. ESA, NASA and JAXA).

First, in Chap. 1, an introduction to satellite data-handling will be given, focusing on the anatomy of a generic spacecraft, paying attention to the internal communication system and its requirement. State-of-the-art solutions will be presented, including an analysis of the already available and future high-speed technological solutions. In Chap. 2, we will present in detail the SpaceFibre standard itself. The relevant ECSS standard will be taken as a reference point. However, the different protocol layers will be presented with a system user perspective, describing protocol mechanism but also available features, points of strength and weaknesses. In Chap. 3, all the different hardware building blocks that can be found in a

SpaceFibre network will be presented, taking as reference example real devices already used by the space community. In particular, the following devices will be presented: a SpaceFibre CoDec, a SpaceFibre router and a SpaceFibre electrical ground segment equipment. Once that the protocol has been introduced and the potential building block of a network has been presented and analysed, the next step in SpaceFibre network development is to check for interoperability of available devices. Although being compatible with the standard already ensures different implementations to be interoperable, at present no conformance checker is available. Therefore, it is considered a good design practice to assess the interoperability of a device. The interoperability test campaign is fundamental in the case that a new endpoint is developed from scratch to assess standard conformance. In Chap. 5, indications are given on how to interconnect different available devices to create a complex SpaceFibre network. A generic satellite onboard data handling network interconnects at least the payload, a control unit and a mass memory, possibly through a routing switch. Usually, this network can be also connected externally to an electric ground segment equipment, for design and debug purposes. Moreover, this hardware can also be simulated and co-simulated as hardware-in-the-loop thanks to advanced simulators. All these aspects will be analysed in Chap. 5, guiding system developers in the understanding and setup of their own SpaceFibre-based system. In Chap. 6, an overview of the technological progress of the SpaceFibre-based system is carried out, in the form of a detailed analysis of the state of the art. It will indicate to technology users the impact of different SpaceFibre CoDecs & routers on the most essential FPGAs and silicon technology. Also, an analysis of the electrical ground segment equipment available in literature will be carried out, to ease the work of future system adopters. Finally, in Chap. 7, we conclude the work, focusing on the future of the protocol. There are also two appendices available, both focusing on a slightly modified version of the SpaceFibre protocol, which may be used in low-cost space missions. The authors would like to acknowledge all the people that worked in the Electronics System Laboratory at the University of Pisa and in the IngeniArs S.r.l. company, especially Alessandro Leoni and Daniele Davalle, who contributed actively to the development of the work presented in this book. We would also like to acknowledge the entire SpaceFibre community, which under the wise guidance of the European Space Agency, has been able to cooperate in the development of this technology.

Pisa, Italy

Pietro Nannipieri
Gianmarco Dinelli
Luca Dello Sterpaio
Antonino Marino
Luca Fanucci

Contents

Acronyms

ACK	ACKnowledge
API	Application Programming Interface
ARINC	Aeronautical Radio INCorporated
ASIC	Application-Specific Integrated Circuit
BC	BroadCast
BE	Best Effort
BER	Bit Error Rate
BFM	Bus Functional Model
BRAM	Block RAM
CAN	Controller Area Network
CDR	Clock Data Recovery
CMOS	Complementary Metal-Oxide-Semiconductor
CoDec	Coder-Decoder
cPCIe	compact Peripheral Component Interconnect express
CRC	Cyclic Redundancy Check
CRD	Current Running Disparity
CSA	Canadian Space Agency
CT	Continuous Traffic
DICE	Dual Interlocked storage Cell
DSP	Digital Signal Processor
DUT	Device Under Test
EBF	End Broadcast Frame
ECSS	European Cooperation for Space Standardization
EDAC	Error Detection and Correction
EDF	End of Data Frame
EEP	End Error Packet
EGSE	Electrical Ground Support Equipment
EO	Earth Observation
EOP	End of Packet
ESA	European Space Agency
FC FSM	Flux-Control FSM

FCT	Flow Control Token
FDIR	Fault Detection Isolation and Recovery
FF	Flip-Flop
FIFO	First-In-First-Out
FPGA	Field Programmable Gate Array
FSM	Finite State Machine
GAR	Group Adaptive Routing
GUI	Graphical User Interfaces
HIL	Hardware-In-the-Loop
ICU	Instrument Control Unit
iLLCW	Inverse Lane Layer Control Word
IN VC	INput Virtual Channel
IP	Intellectual Property
ISO	International Organization for Standardization
JAXA	Japan Aerospace Exploration Agency
LCROSS	Lunar CRater Observation and Sensing Satellite
LET	Linear Energy Transfer
LLCW	Lane Layer Control Word
LSS	Lowest Significant Symbol
LUT	Look Up Tables
MAC	Medium Access Controller
MCU	Main Control Unit
MIB	Management Information Base
ML-FSM	Multi-Lane FSM
MMFU	Mass Memory and Formatting Unit
MSS	Most Significant Symbol
NASA	National Aeronautics and Space Administration
NDCP	Network Discovery and Configuration Protocol
NGSIS	Next Generation Spacecraft Interconnect Standard
OBC	On-Board Computer
OS	Operative System
OSI	Open Systems Interconnection
OUT BC	OUTput Broadcast
OUT VC	OUTput Virtual Channel
PCIe	Peripheral Component Interconnect express
PL	Programmable Logic
PLL	Phase-Locked Loop
QoS	Quality of Service
RAM	Random Access Memory
RC	Rate-Constrained
RefArc	Reference Architecture
RMAP	Remote Memory Access Protocol
RR	Round Robin
RRArbiter	Round Robin Arbiter
R-SpFi	R-SpaceFibre

RT	Reliable Traffic
RTL	Register Transfer Level
SAE	Society of Automotive Engineers
SAR	Synthetic Aperture Radar
SAVOIR	Space AVionics Open Interface aRchitecture
SBF	Start Broadcast Frame
SEE	Single Event Effect
SEL	Single Event Latch-up
SerDes	Serialiser-Deserialiser
SEU	Single Event Upset
SHINe	Simulator for HIgh-speed Network
SOC	System on Chip
SpFi	SpaceFibre
SWIP	Switching Block IP
TB	TestBench
TCP	Transmission Control Protocol
TID	Total Ionization Dose
TMR	Triple Modular Redundancy
TT	Time-Triggered
TTEthernet	Time-Triggered Ethernet
UDP	User Datagram Protocol
UUT	Unit Under Test
UVM	Universal Verification Methodology
VC	Virtual Channel
VHDL	Very high-speed integrated circuits Hardware Description Language
VN	Virtual Network
WISM	Word Identification State Machine

Chapter 1
Introduction to Satellite on-Board Data-Handling

1.1 Anatomy of a Spacecraft and Requirement Analysis

1.1.1 Earth Observation and High-Resolution Payloads

Earth Observation (EO) satellites are an extraordinary tool for collecting information about our planet via remote sensing technologies, helping us to understand how the Earth system responds to natural and human-induced changes. Starting from the '60, EO satellites found applications in several different fields, such as environment monitoring [44], weather prediction [47], land management and cover [61], maritime surveillance [71], agriculture [13], food security [4], disaster monitoring and managing [19] and homeland security [16]. Today, EO is not only a powerful science instrument, but it has also relevant economic, environmental and societal impacts [40]. The United Nation recognised the importance of EO for the achievement of the 17 Sustainable Development Goals program [2] that aims at improving prosperity for people and the planet, building a more sustainable world [80].

The Sputnik 1 can be considered the very first EO satellite because it sent back to Earth the radio signals that Russian scientists used for studying the composition of the Ionosphere [74]. The Landsat mission was the first to downlink a consistent number of images: launched starting from 1972, the Landsat satellites send back to Earth approximately 2 million images [40]. Since then, more and more satellites for EO applications were launched, and Fig. 1.1 shows the number of active EO missions according to the World Metrological Organisation (WMO) [82]. Historically, the United States, Russia, Italy, France and Germany are the leaders in the EO satellite launches, but other states such as China, India, Brazil, Canada, Australia, Nigeria funded their EO missions in the recent past [36]. Approximately 50 countries are now investing in EO programs, and the EO market is expected to continue to grow in the next years [15]. At the same time, EO is rapidly changing as the result of the advances in digital technologies and sensor

P. Nannipieri et al., *Next-Generation High-Speed Satellite Interconnect*,
https://doi.org/10.1007/978-3-030-77044-0_1

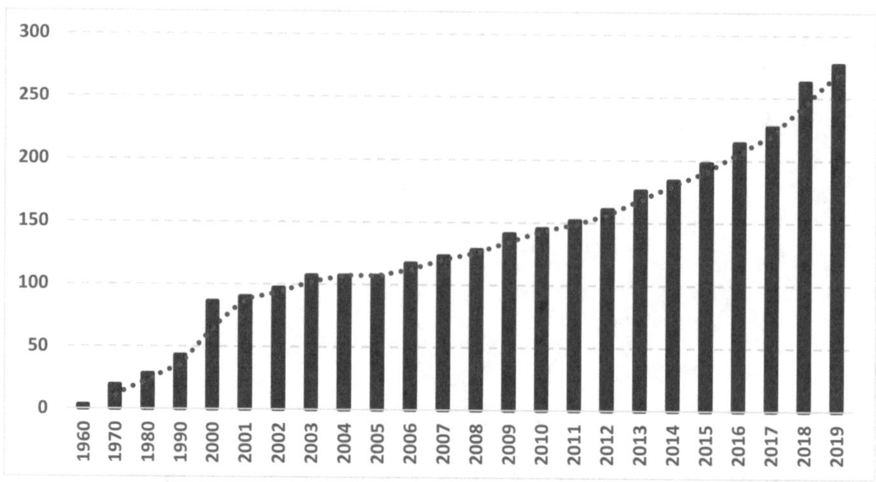

Fig. 1.1 Active EO satellite missions per year according to the WMO

resolution [7, 53]. There is a continuous search for payloads that can offer images with a higher resolution both in the commercial and defence markets [15]. As result, many recently launched and future planned missions mount high-resolution instrumentation, such as Synthetic Aperture Radars (SARs) and hyperspectral images. SARs are commonly mounted on aeroplanes or spacecraft and produce 2-D images or 3-D reconstruction of objects, exploiting the motion of a pulsed radar over the region to be studied [57]. A SAR receives the echoes of the backscattered signal, whose amplitude and phase depend on the physical and electrical features of the object observed. SARs provide high-resolution images independently from daylight, cloud coverage and weather conditions [31, 64], and their applications range from geoscience to climate change research, from 2-D and 3-D mapping to security-related applications [57]. Hyperspectral imagers collect data of hundreds of narrow contiguous spectral bands in order to identify different materials from their spectral signature [54]. For this reason, hyperspectral images are often referred as "cubes". Hyperspectral imagery is exploited for target detection, material mapping and material identification [73]. Figure 1.2a and b show examples of SAR and hyperspectral images, respectively.

Some relevant examples of missions that recently involved this class of payloads are Envisat [51], Sentinel-1 [49], Sentinel-2 [18], Sentinel-3 [17] and PRISMA [50]. CHIME [62], NISAR [43], HyspIRI [8], EnMAP [34] and FLEX [69] are examples of National Aeronautics and Space Administration (NASA) and European Space Agency (ESA) missions mounting SARs that will be launched in the near future.

Besides high-profile missions, small satellites and, in particular, CubeSats are begging to play an important role in EO [10, 72]. CubeSat is a class of nanosatellite (with a mass between 1 and 10 Kg) standardised by the California Polytechnic State University in 1999. They are made up of $10 \times 10 \times 10$ cm units (1U) with a maximum

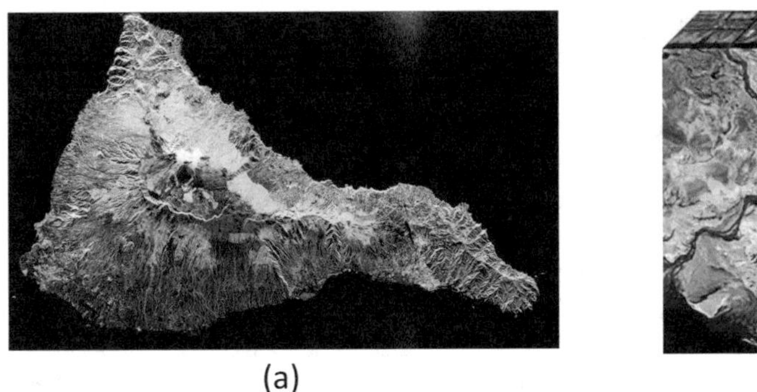

(a)

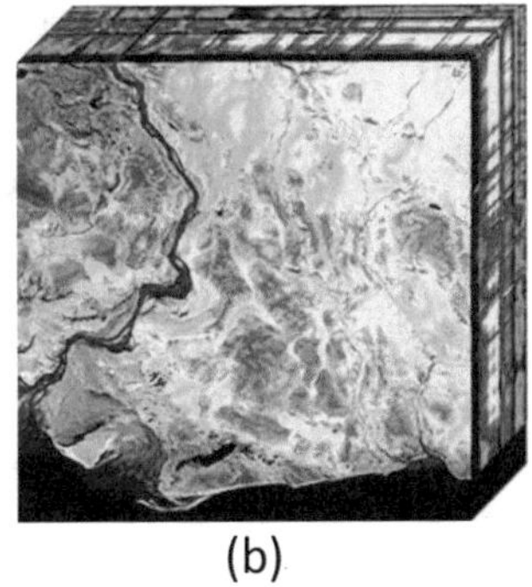

(b)

Fig. 1.2 In (**a**), an example of a SAR Image [59]. In (**b**), a hyperspectral cube [60]

weight of 1.33 Kg [14] that could be assembled to compose larger satellites (e.g. 2U, 3U, 6U and 12U). CubeSats, initially developed with educational purposes, despite their limitation in terms of mass and volume are emerging as important technological platforms [81], especially for EO applications, representing a cost-effective and fast-to-launch solution [66].

CubeSats have a limited cost, ranging from few tens of thousands to a few million thousand Euro per unit [15, 81] and can also be part of large constellations with the potential of achieving comparable or even better performances than traditional spacecraft [5, 72]. Currently, several CubeSat missions already plan to mount SAR, such as CIRES [83], Phisat-1 [20], Capella 1 [9] and ICEYE [37], and hyperspectral imagers, such as Intuition-1 [45], HyperCube [46] and Waypoint 1 [75].

1.1.2 Satellites and the On-Board Data-Handling Sub-system

A spacecraft is a machine designed to fly in outer space. It may be used for various purposes, including Earth Observation (EO), space exploration, communication and many others. Spacecraft is often remotely operated, except for few manned missions. They shall be extremely safe to operate, both for ethical (e.g. manned missions) and economic/strategic reasons. The design of a spacecraft requires huge engineering efforts: it is a complex system that has to withstand intense mechanical and thermal stresses, with strict requirements in terms of reliability. This book does not focus at all on the mechanical and thermal requirements involved in the design of a spacecraft; indeed we intend to focus on its electronic system, and in particular, to the mechanisms exploited for collecting and elaborate data: in our preliminary assumption, at least from an electronic point of view, a spacecraft is a system acquiring data through sensors, receiving commands from a remote user (ground) and sending back the acquired data to the remote user. Although in modern spacecraft this is not always the case, we can take it as a solid baseline. The data processing requirements of a satellite are mostly directly related to the electronics of the spacecraft itself, and considering how the world of electronics has evolved in the past decades, we cannot think that the on-board data-handling systems for spacecraft have not changed deeply during history [79]. In the following, we will refer to avionics as the electronic system of a spacecraft. It includes communication, navigation and on-board data-handling. An avionic system is generally composed of the payload, which is the equipment performing mission-specific tasks, and the platform, which performs all satellite operational tasks, including data processing, data storing, navigation and telemetry. It is not our aim to fully detail the complex avionics architecture of a spacecraft, also because different missions have different on-board communication architectures (for detailed analysis on digital avionics architectures please refer to [29]). However, such a complex system shall have several communication links between separate modules, and within the modules as well. In Fig. 1.3 a typical earth observation/scientific spacecraft network topology is systematised. It represents a general high-speed communication architecture for space applications. A spacecraft may mount several instruments, each one producing a significant amount of data that shall be processed: each instrument is usually connected with an electronics system that pre-processes the acquired data and operates the instrument through the Instrument Control Unit (ICU). Data and commands communication lines are then sent to the Mass Memory and Formatting Unit (MMFU), where data is distributed through a routing switch to the other components of the avionics: a mass memory to save data waiting to be transmitted to earth, a Main Control Unit (MCU) operating the entire system and the downlink formatter, for the communication with the ground station (Earth). Redundancy is another important concept displayed in Fig. 1.3: it is common in spacecraft to have redundant components, and this also affects the communication system, or at least its command & control section. Redundancy is introduced to improve reliability: a component may have a failure, a cable may be damaged during the launch of the spacecraft, and this shall not invalidate the entire mission. Of course, this redundancy has disadvantages: the on-board data-handling system gets more complex, there are more cables and components, which means a heavier system and also more expensive.

Now that we had a quick view of a spacecraft avionics block scheme, we are also ready to introduce better the constraints that such systems shall usually withstand. The first requirements for spacecraft come directly from the operative environment: it is known that semiconductors are susceptible to a fault (definitive or not) when exposed to a certain radiation dose [6, 63, 84]; if the circuit is supposed to operate in a harsh environment, such as outer space, where the exposure dose is much higher, an electronic system must be properly designed to operate safely. This means that spacecraft avionics shall be built upon specific silicon technologies, able to cope with the radiation level. Redundancy is a key concept in space avionics design, due to the high cost of spacecraft and the intrinsic difficulty in operating maintenance on them. Therefore, several solutions are usually adopted to strengthen the fault tolerance of the spacecraft avionics system, e.g. each communication link can be doubled. Carrying objects in space has a cost, which is also related to their weight [41]. This is valid also for cables: fewer cables mean lower harnesses, easier

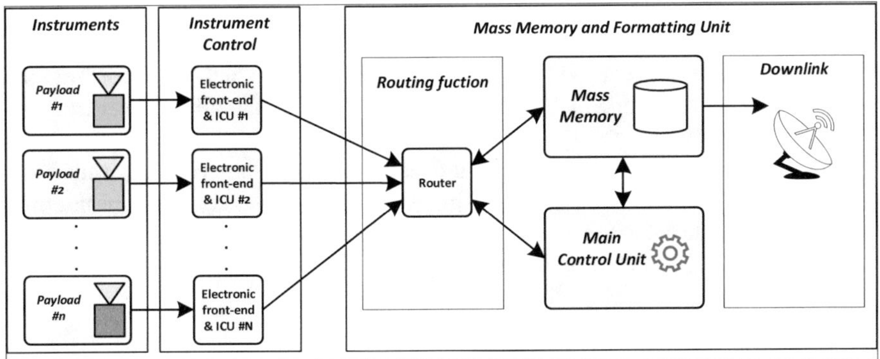

Fig. 1.3 Generic spacecraft network topology

integration and reduced costs. Cable reduction can be achieved by lowering the number of physical communication links and, as a future perspective, by replacing classical copper wires with optical fibre. The need to reduce the weight could also push, in the foreseeable future, towards the use of the same network infrastructure for both science data (not time-critical, high data rate) and command & control data (time-critical, low data rate). Networks will, therefore, need to be able to offer adequate Quality of Service (QoS), to allow different traffic classes to coexist on the same physical medium and protocol infrastructure [58]. It is possible, especially with the use of Synthetic Aperture Radars (SARs) and hyperspectral imagers, e.g. European Space Agency (ESA) upcoming missions (FLEX [69], BIOMASS [11], Sentinel-2 [28], NISAR [43], HyspIRI [8]), that very high data rate is required in spacecraft on-board communications: a large number of instrument can be hosted on the same vehicle, sharing resources and communicating through various sub-system. At present time a data transmission rate in the order of several Gbps is required and the trend is currently growing. The combination of the aforementioned requirements results in several options for data-handling network topology: complex systems may require to include a router, to share and multiplex a single (possibly high-speed) communication medium. All these constraints, combined with the high reliability and efficiency, typically requested in the space field, do not allow designers to adopt general-purpose communication solutions. In particular, the significant growth of data rate requirement led different institutes, agencies and companies to start working on new on-board data-handling communication protocols to meet this request. A set of attributes and features will be used in this book to characterise on-board data-handling systems. In the following, we introduce the key concepts behind these features.

- *Quality of Service*: an on-board data handing protocol providing this service can multiplex and schedule several communication channels on the same physical link, according to specific requirements, e.g. priority and bandwidth allocation.

- *Fault Detection Isolation and Recovery (FDIR)*: an on-board data handing protocol providing this service can detect faults occurring on the network, which are most of the time caused by space natural radiations, isolate the error from the rest of the data stream which is still considered valid and finally recover from the error asking for a data retry to the far-end of the communication.
- *Determinism*: a deterministic on-board data handing protocol can set up communication with bounded latency. This is usually a mandatory feature not only for time synchronisation mechanism across the spacecraft but also for sensor actuators (e.g. parachute opening).

1.1.3 Space Avionics Open Interface Architecture

After a brief introduction of the main building blocks of a spacecraft, we need to introduce a more detailed and standardised satellite avionic system. Even though it is not the aim of this work to provide in details the description of spacecraft avionics, it is important to understand the complexity of the topic and the high-level requirements, which will lead then to the definition of communication protocols. The Space AVionics Open Interface aRchitecture (SAVOIR) is an initiative coordinated by ESA, in collaboration with the space industry, to federate the space avionics community and to work together to improve the way the European space community builds avionics sub-systems. Its main objectives are to help the standardisation process of space avionics, to reduce risks, costs and promote system re-usability. One of the first documents produced and officially released by SAVOIR is the On-board Communication System Requirement Document [76]. Based on the SAVOIR framework, the aim of [76] is to define an avionics functional reference architecture, to meet the needs of various mission domains and to provide generic requirements related to current and future avionics communication needs. It can, therefore, be considered as a common-core of requirements that are expected to be relevant to a sizeable range of future missions. The approach is protocol independent, representing an ideal baseline for the work presented in this book. The document [76] suggests a series of definitions, reported in the following paragraphs, that will be useful throughout our work.

Communication System
An (on-board) communication system denotes any on-board architecture that permits to perform information transmission on-board a spacecraft. According to the chosen technology, redundancy schema and physical architecture, a communication system can be implemented as:

- A point-to-point link: a dedicated communication connection between two communication endpoints or nodes.
- A bus: communication system that transfers data between endpoints sharing a common communication infrastructure.
- A network: a communication infrastructure involving two or more nodes or endpoints, connected employing point-to-point links and routing switches.

A communication system is to be understood as the union of several endpoints and routers or switches, bus or other interconnection topology that interconnects them, designed to offer a set of properties and services (bandwidth, communication delays, communication delays variation, packet loss statistics, data streams segregation, priority management) that can be guaranteed and referred to as QoS.

Jitter
The jitter is defined as a timespan around an expected time of action within which the action is performed. The action is e.g. a sensor acquisition or an actuator command, or a message delivery.

End-to-End Latency
The end-to-end latency is the total timespan necessary from when a client application of the communication system wants to submit a message for transmission until the message is received at the destination end, ready to be used.

Message Transmission Latency
The message transmission latency is the timespan necessary to send a complete and formatted message from the source endpoint to the destination endpoint. It measures the time interval between the moment when the message is ready to be transmitted on the communication system until its complete reception at the destination end. The transmission latency is therefore influenced by the effective usage of the communication system in terms of message traffic and possible interference with the message of interest. Other important factors are the routing or bridging element in the case of non-point-to-point communication and the QoS implementation (acknowledge management, handshake, re-emit, etc.).

Virtual Channel
The Virtual Channel (VC) is a technique for multiple on-board packet sources (application processes) to share the finite capacity of the physical link through multiplexing. Each virtual channel is identified by a unique virtual channel identifier. Frames from different virtual channels are multiplexed together on a master channel corresponding to a physical link.

SAVOIR tries to be as agnostic as possible for what concerns the interconnection topology, although the two major topologies that are a candidate to support these requirements are the bus and network topology. Within SAVOIR output products, general and functional reference architectures for avionics systems have been created, as shown in Figs. 1.4 and 1.5.

Although Fig. 1.4 is quite complex, it is straightforward to understand the complexity of a system build on top of these specifications. Even if we do not consider in our analysis the *Applications* and *Execution Platform* sections, which are implemented in the software stack, the number of hardware building blocks is remarkable; the following hardware is included in the SAVOIR reference avionics system:

- *Processing Function*: It includes all the control feature of a spacecraft, such as telemetry interfaces, payload and platform command & control, on-board time distribution.

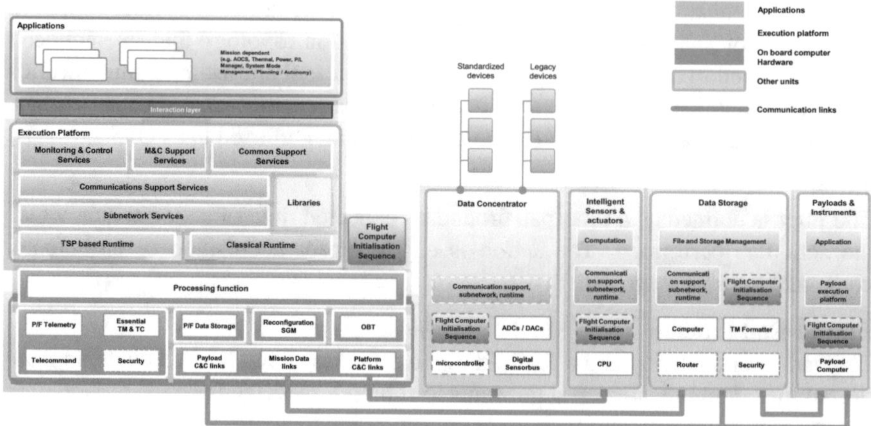

Fig. 1.4 SAVOIR reference architecture [78]

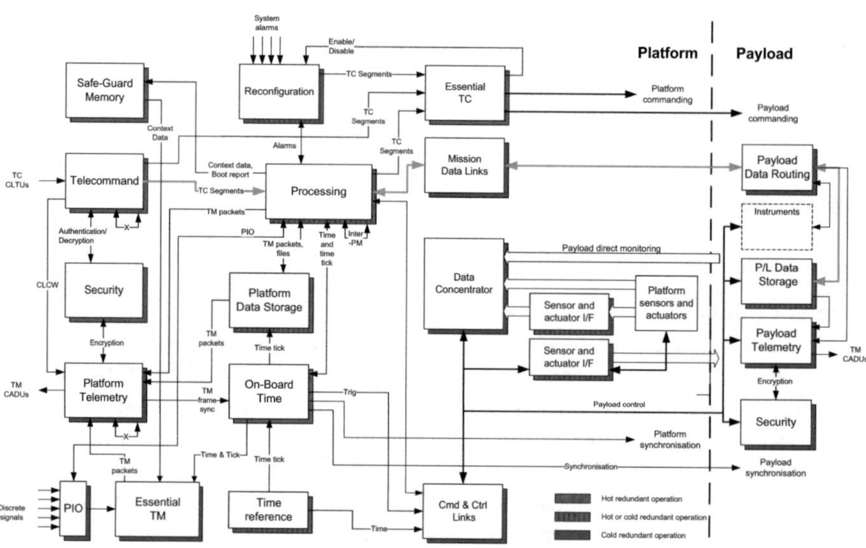

Fig. 1.5 SAVOIR functional reference architecture [78]

- *Data Concentrator and Intelligent Sensor & Actuators*: Sensors and peripheral devices data generation, collection and distribution.
- *Data Storage*: Storage mechanism for both spacecraft housekeeping and payload generated data.
- *Payload & Instruments*: Biggest share of generated data within the spacecraft, which has to be acquired, pre-processed and distributed within the spacecraft, and also sent back to the ground.

The SAVOIR functional Reference Architecture (RefArc) identifies two on-board communication functions: The *Command and Control Links* and the *Mission Data-Links*. The command and control links function concerns the capability of commanding the whole avionics from the processing function. Additionally, it permits to perform command and control of payload functions, such as *payload data storage*, *payload telemetry* and *payload-related security function*. The function is also related to on-board time, which is distributed via the command and control links. The *mission Data-Links* function concerns the capability of the processing function to perform routing of payload-related data traffic. This is important as it concerns at the same time payload telemetry and payload data storage. The platform data storage provides the means to store spacecraft housekeeping data and operation data (mission timeline, spacecraft on-board control procedures, etc.); it is usually implemented (typically in a hot redundant configuration) in modules or boards of the On-Board Computer (OBC) unit communicating with the Processor and the platform telemetry functions. The payload data storage acquires and stores payload mission data (e.g. from payload instruments); it can be physically implemented in a single unit integrated with the payload transmitter, or implemented as a self-standing storage unit or even integrated with the platform data storage (e.g. both included in the OBC).

If we take into account the engineering of the on-board data-handling network of a spacecraft, we can observe the following: SAVOIR reference architecture requires a great number of devices to be interconnected, with very different specification, often coming to trade-off, e.g. bandwidth Vs time criticality. It is therefore fundamental to have adequate protocol infrastructure to address the communication needs of modern spacecraft.

1.2 State-of-the-Art Solutions

We already mentioned that several strict requirements usually apply to the on-board data-handling sub-system of a spacecraft: bandwidth shall be large enough to support modern payloads, weigh shall be minimised, redundancy shall be guaranteed, radiation tolerance and/or radiation hardening shall be considered and the system cost is getting more and more an important parameter to be minimised. These requirements push towards the use of the same network infrastructure for both science data (not time-critical, high data rate) and command & control data (time-critical, low data rate), which shall offer adequate QoS support to allow both the traffic classes to coexist. Such a network should be able to support:

- **Best effort traffic**, to carry high-speed scientific data generated by the payloads. The deterministic delivery of all the data and their correctness is not a critical point for this kind of traffic.

- **Synchronous time-critical traffic**, usually small packets generated by control instruments with a predefined schedule. They usually need a deterministic and time-bounded delivery to the destination.
- **Asynchronous time-critical traffic**, generated by control logic when an event occurs, for example, upon the detection of an error.

All these constraints, combined with the high reliability and efficiency typically requested in the space field, do not induce designers to adopt general-purpose communication solutions. However, the state-of-the-art solutions do not completely answer the needs of system designers. Two orders of problems particularly affect the current design of data-handling systems: command and control networks and data-handling networks are currently segregated, without any plan to be merged by the space community; even if the technology would allow it, the risk reduction requirement is against this possibility, at least in the foreseeable future. This empowers resiliency of obsolete command & control protocol, such as MIL1553 [48]. To minimise risks, it would also be necessary to adopt standardised and open protocols. However, currently, no standardised open protocol is adopted in missions requiring consistent data rate (above the Gbps). Therefore, for high-speed data-handling, either not standardised solutions such as WizardLink are adopted or more standardised interfaces (e.g. SpaceWire) are used in parallel, with drawbacks in terms of resource, power consumption, harness and performances. In the following, an overview of how those protocols handle current system requirements is given.

1.2.1 SpaceWire and WizardLink for Data Transfer

SpaceWire

SpaceWire is a satellite on-board communication protocol developed under the supervision of the ESA by the University of Dundee [65]. It is an open protocol, standardised by European Cooperation for Space Standardisation (ECSS, ECSS-E-ST-50-12C) [26] in 2003. The protocol has been recently revised in 2019, thanks to the effort of the worldwide space community, with contribution from all the major space agencies such ESA, National Aeronautics and Space Administration (NASA), Japan Aerospace Exploration Agency (JAXA) and ROSCOSMOS. SpaceWire has been designed to be a low error rate line, with low resources utilisation (5–8 Kgates [56]). The communication line is full-duplex. The protocol can reach theoretical speeds of 400 Mbps in on-board applications, making it a valid solution in the past years for payload data transfer. SpW is currently exploited in several high-profile satellite missions of agencies such as ESA, NASA, JAXA, Roscosmos and Canadian Space Agency(CSA) [3]. A partial list of relevant missions includes NASA Lunar Reconnaissance Orbiter, Lunar CRater Observation and Sensing Satellite (LCROSS), Swift and Magnetospheric Multiscale, the NASA/ESA/CSA James Webb Space Telescope, ESA Gaia, ESA/Roscosmos ExoMars, ESA/JAXA BepiColombo and JAXA Hayabusa2 and Astro-H. Such a great heritage from previous missions makes it a highly reliable solution. A wide set of solutions of Coder-Decoders (CoDecs) and routers are available on the market, in both Application-Specific Integrated Circuit (ASIC) and Field Programmable Gate Array (FPGA) [27], and also a wide range of support equipment is available [52]. The protocol itself has been designed for easier integration and to improve and promote equipment reuse across different missions, making it a competitive choice in terms of costs.

However, SpaceWire has also several disadvantages: it cannot be considered a deterministic protocol, with the presence of routing switches inside the network creating even greater time uncertainty.

It does not provide time determinism on its own for data packets. Indeed, it does not offer any mechanism to prioritise a critical message over the normal data stream. SpaceWire point-to-point links are not fault-tolerant and do not foresee any data retry mechanism; SpaceWire is not fault-tolerant even at the network level, because of the wormhole forwarding scheme in the switches that, combined with the undetermined length of the packets, could lead to unbounded delays. In a network using the wormhole forwarding scheme, when a packet starts to flow through one of the output router's ports, that port is locked and is not usable by any other input port until the packet is completely transmitted. In case the packet is very long or the source node pauses the transmission, the other nodes in the network suffer from large (and possibly unbounded) delays. SpaceWire does not provide reliable service: packets are not protected through a Cyclic Redundancy Check (CRC) by the standard itself, even if typical upper layer protocols do it, such as the Remote Memory Access Protocol (RMAP) [22]. This means that some errors can pass undetected. In case an error on the link is detected, SpaceWire truncates the current packet signalling that an error has occurred, and it tries to reinitialise the link. SpaceWire does not support on its own any QoS. This means that it is not possible to differentiate the traffic depending on the application requirements.

WizardLink

WizardLink is a communication protocol developed by Texas Instrument which also produces the chip-set necessary for the communication (TLK2711) [85]. It is a lower layer protocol, not standardised, adopted in several ESA Earth observation satellites, such as Sentinel-1, Sentinel-2 and MTG, due to its high data rate, up to 2.5 Gbps. WizardLink has been used and designed only to implement high-speed payload to platform communication. The protocol features 8B/10B data encoding and does not include any higher layer features, such as FDIR or link initialisation handshake: these functionalities have to be implemented by the user, leading to longer integration time and lower reliability. The technology is designed and developed in the form of the product only in the USA, making it critical for European applications to rely on it. Anyway, products related to this technology are available as rad hardened IC [86], ready to be used. The TLK2711 has been used as a lower Lane layer and Serialiser–Deserialiser (SerDes) in many SpaceFibre CoDec demonstrations, as shown in [33]; indeed, the SpaceFibre standard itself has been explicitly designed to be compatible with WizardLink. WizardLink does not provide any kind of determinism, neither point-to-point nor end-to-end, and does not provide any reliability support for the network. Also, no QoS is provided.

1.2.2 MIL1553 or CAN for Command and Control

The **Mil-Std-1553B** (or Milbus) is a communication protocol standard adopted in its B version in 1978 by the US air-force. It defines characteristics of a serial multiplex data bus, with a set of electrical, mechanical and functional requirements. The standard was born for military uses, but its field of application has been soon extended to other fields such as aerospace. It has been adopted by the main space agencies (e.g. NASA and ESA), leading to an organisation-specific version of the standard (e.g. ECSS-E-50-13C [21] for ESA). The **CAN 2.0** (Controller Area Network) is a standard published by Bosch in 1991, evolved from the original CAN standard, and has then been integrated by ECSS for space use in the ECSS-E-ST-50-15C [24] standards, introducing redundancy at architecture and software level. The protocol itself consists of a balanced differential bus. It can reach a data rate of 1 Mbps or 8 Mbps with CAN FO, making it not applicable for a modern payload to the platform communications line. Anyway, for its large consumer heritage, it is still a robust option for command and control communication lines.

1.3 Candidate Solution for Next-Generation Avionics System

SpaceWire (and Its Extensions)

The fact that SpaceWire is not a deterministic protocol itself was not acceptable in some application scenarios where determinism was fundamental. However, being SpaceWire a low-level communication protocol, it is possible to exploit its existing infrastructure to add new features, in the form of upper layers protocols. The SpaceWire user community pushed for the release of an additional standard named SpaceWire-D, which implements deterministic features on top of the SpaceWire protocol.

The SpaceWire-D protocol has been developed to overcome the lack of time determinism. It is an upper layer protocol relying on RMAP [23] that can add the time determinism to a SpaceWire network using a time-code based ad-hoc synchronisation protocol. SpaceWire-D is usually implemented as a software layer in the nodes due to its complexity. It is important to note that, even if SpaceWire-D is used, the network remains not fault-tolerant and a failing node can still cause undetermined delays. Also, an extension named SpaceWire-R, which provides reliable data transfer at the network level, is currently under development.

TTEthernet

Time-Triggered Ethernet (TTEthernet) is an extension of the IEEE 802.3 Ethernet standard [38] that provides deterministic communication with real-time guarantees and predictable maximum delay and jitter. It is a suitable solution both for safety-critical systems that require knowing the temporal behaviour of a data packet but can also include the traffic of non-critical applications. TTEthernet combines three standards that provide support for three different traffic classes:

- the SAE AS6802: the Society of Automotive Engineers (SAE) AS6802 standard [70] supports Time-Triggered (TT) messages, which are used to ensure the determinism of the communication process. TT messages are sent over a TTEthernet network at predetermined times and have higher precedence than the other traffic classes. Thus, two TTEthernet senders will never send a TT message to the same receiver, so that packets cannot be queued in network switches. Knowing in advance the maximum delay and jitter of TT messages, TT traffic is the suitable solution for real-time systems and safety-critical applications, requiring determinism.
- the Aeronautical Radio INCorporated (ARINC) 664P7 standard [39] provides support for Rate-Constrained (RC) Traffic. This class of traffic does not have a strict timing requirement, requiring only an upper bound of the transmission delay that can be guaranteed since each TTEthernet node has a dedicated bandwidth to RC traffic. RC messages find applications in scenarios that do not require strict real-time and determinism constraints. Typically, they can be employed in safety-critical application with moderate requirements in terms of latency and jitter.
- the IEEE802.3-2005 standard [38], supporting the Best Effort (BE) traffic, has not guaranteed performances in terms of data rate. BE traffic class is not deterministic, because TT and RC messages have a higher priority than BE messages, which exploit the residual available bandwidth.

TT and RC messages are framed according to the specification of ARINC 664P7 standard, while BE messages are embedded in the traditional Ethernet frame. Available TTEthernet products support 100 Mbps and 1 Gbps data rates since they are compatible with Gigabit Ethernet. The structure of the TTEtherent allows data coming from the payload to be transmitted together with control communication, guaranteeing a very flexible architecture to be implemented. The protocol provides fault detection and isolation, but no retransmission and built-in fault-tolerant network synchronisation. This technology, for its compatibility with the regular Ethernet, is oriented towards launchers, manned missions and space exploration missions, but its applicability for telecom and earth observation missions is under investigation. The technology is already used in the Orion program and should be implemented in the Ariane 6 program as well [1]. Possible drawbacks of this standard are higher mass, power consumption and complexity, considering that TTE is born from the combination of three different protocols (Time-Triggered Protocol and Gigabit Ethernet) [35]. Also, the protocol itself is not an open solution, which is not the best option in terms of reliability and technology dependency.

TTEthernet relies on Gigabit Ethernet, thus its current maximum data rate is 1 Gbps, net of the encoding. As standard Ethernet, the link efficiency of TTEthernet greatly depends on the payload size transmitted in each frame. TTEthernet does not have additional traffic like acknowledgements or Flow Control Tokens (FCTs). It can guarantee deterministic packet delivery and it uses a master-slave synchronisation protocol to synchronise all the nodes in the network. TTEthernet is also fault-tolerant, allowing the switches to drop inconsistent messages from incoming links. For what concerns reliability, TTEthernet is not reliable. Ethernet itself does not provide any mechanism of retransmission in case of errors and any error detection and recovery is demanded to upper-layer protocols. TTEthernet does not even implement a flow control mechanism. Depending on the kind of application, both these services can be added using common protocols such as TCP/IP, which can be easily integrated into a TTEthernet network. For what concern redundancy, TTEthernet supports a fault-tolerant system including redundant network paths. If a path is recognised as faulty, the communication process can continue on redundant paths without affecting system operations. A TTEthernet end can support 1,2 or 3 redundant channel configurations.

The RT traffic is used for packets having not too strict temporal requirements but that still requires guaranteed bandwidth. Each node in the network has a reserved bandwidth for the transmission of RC messages, so it is possible to calculate the upper bound for latency and jitter. However, these messages are not sent accordingly to a system-wide synchronisation mechanism, thus multiple RC packets may be sent to the same destination at the same time, increasing latency. Finally, BE traffic can be used for any left non-time-critical application. It offers the same service as a normal Ethernet connection, so there are no guarantees both on the maximum delay and on the correct delivery (packets can be lost due to overflow in buffers). BE traffic has the lowest priority among the three classes of QoS. In any case, the classic upper-layer protocols such as the Internet Protocol (IP) [67], User Datagram Protocol (UDP) [77] and Transmission Control Protocol (TCP) [68] can be used, as shown in Fig. 1.6.

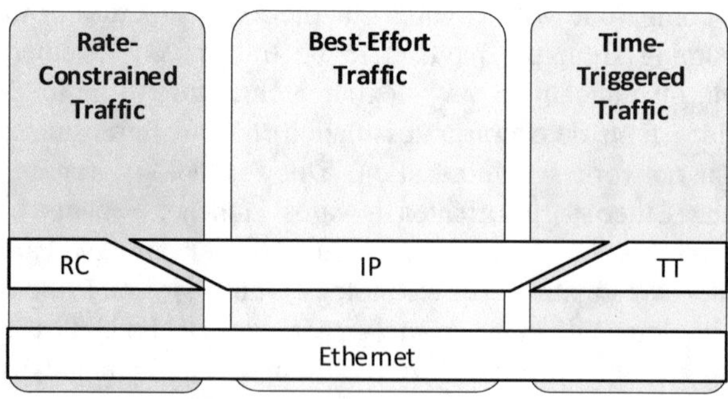

Fig. 1.6 TTEthernet stack and services

RapidIO 3.1

RapidIO 3.1 has been released in 2014, from a collaboration between the RapidIO trade association and the Next-Generation Spacecraft Interconnect Standard (NGSIS) [32]. Its not-deprecated version is Serial RapidIO. It is a high-performance packet-switched interconnect technology, supporting real-time command and data-handling, signal processing and high-reliability networking and communication [35]. It can reach data rates of 6.25 Gbps, reducing power consumption including asymmetric links. Indeed, payload to platform communication is almost unidirectional in most cases since usually, the instrument produces a significant amount of data which are then unidirectionally sent to the elaboration chain. RapidIO implements fault tolerance mechanism, hardware-based error recovery and flow arbitration [42]. It has been on the consumer market for more than 10 years, but at the time of writing, there are no space-qualified products up to the author's knowledge. RapidIO 3.1 supports a maximum data rate of 6.25 Gbps, with the option to use multiple lanes to increase the bandwidth, and it can run on both copper wire and optical fibre.

RapidIO QoS provides supports VCs (up to 9), implementing two different traffic classes:

- Reliable Traffic (RC) that guarantees reliable communication, providing for corrupted packets retransmission and preventing overflow of receiving buffers.
- Continuous Traffic (CT) for not reliable communication allows reducing the overall latency that can be increased by the retransmission process.

In reliable mode, the packets are protected by an end-to-end CRC. When a packet is correctly received, the destination node sends an acknowledge back to the transmitter that can free its buffer. In case of error, the recipient notifies the transmitter that tries to send the packet again. Packets longer than 80 bytes have an intermediate CRC to minimise latency in case of retransmission. In continuous mode, there is no retransmission in case of buffer overflow at the recipient or a physical error on a link. This mode can be used to reduce the jitter and the delay in the transmission of packets. RapidIO implements the concept of QoS through the use of up to 9 VCs and priority levels. The implementation of VCs is optional, provided that at least VC 0 is implemented. VCs from 1 to 8 can be used for bandwidth reservation: it is possible to assign a different expected bandwidth to each one, obtaining a guaranteed link utilisation for the application. VC 0 instead has always precedence over the other channels, and it is the only one allowing to specify packet priority. A packet with higher priority must take precedence over the others, so a reordering mechanism for the buffer is needed. RapidIO supports 4 levels of priority, but it must be taken into account that, because of the protocol architecture, a response message must always have a higher priority to avoid deadlocks. In RapidIO it is possible to disable both retransmission and/or flow control mechanisms, improving the link efficiency, with the only exception of VC 0 that shall work in RT mode. Optionally, switches can implement a time-out mechanism to discard packets that are taking too much time to be dispatched. A RapidIO network can be made fault-tolerant taking extra care of the network configuration. The protocol also offers an isolation mechanism to avoid congestion on links. Regarding the flow control mechanism, RapidIO offers two modes as well. The first is the receiver-based flow control: the transmitter does not have any information on the recipient's buffers status. In case the recipient cannot accept the data, it asks for retransmission (that will take place only in reliable mode). The other mode is the transmitter-based flow control: the receiver side continuously notifies the transmitter about the available space in its buffers and data are sent accordingly to avoid overflows.

SpaceFibre

SpaceFibre is a communication protocol developed by the University of Dundee for ESA [55], born as an evolution of SpaceWire, whose standard has been released in May 2019 after a long review process. It can provide a data rate of several Gbps per communication lane, depending on the transceiver technology (up to 16 lanes operating in parallel), which is even greater than the WizardLink data rate, overcoming its limitations. It can run both on copper cables and optical fibre, making it possible to have a further mass reduction of the spacecraft. The innovative QoS mechanism included in SpaceFibre provides concurrent bandwidth reservations, priority and scheduled QoS. FDIR techniques are integrated, improving system reliability and lowering integration time: error containment in VCs and frames has been introduced, together with transparent recovery from transient errors, galvanic isolation and "babbling idiot" protection. This relieves the upper layers to check the data integrity, hence, reducing both development and system validation time and cost. At the present moment there are several FPGA implementations, and ASICs are under development [12, 56].

SpaceFibre is also backward compatible at packet level with SpaceWire equipment: this easily enables interconnection of SpaceWire devices into a SpaceFibre network, getting the QoS and FDIR advantages of SpaceFibre. SpaceFibre has its network layer, and also a transaction layer not completely defined yet. SpaceFibre can easily guarantee the time determinism and the fault tolerance of packets belonging to different Virtual Networks (VNs). In case the network is not small enough to use a different VN for each source node, attention must be paid to configure the nodes so that they will never compete. A possible solution is to configure the source nodes to transmit in different timeslots and then multiplex all data streams in the same VN at the router level. Also, the implementation of a fault isolation mechanism needs some extra limitations to the standard, such as a limitation on maximum packet size or segmentation of long packets. In SpaceFibre, retransmission in case of errors cannot be disabled, so extra care must be taken to cope with this possible source of non-determinism (for example, forcing the termination of packets taking more time than allowed to be transmitted). SpaceFibre implements reliability through its FDIR mechanism. All the data frames (a segment of 64 data words) and control words are protected by a CRC that is evaluated at each hop in the network. Differently from RapidIO, it is directly the next hop that acknowledges the received data; hence, the transmitter can free its buffer earlier, without waiting for the packet to reach the final destination. SpaceFibre also provides a token-based flow control mechanism that avoids overrunning the recipient resources. In case of a too high error rate, the link is reinitialised. SpaceFibre does not allow to disable the retransmission, but it normally takes less than 1 us. SpaceFibre provides a very flexible and complete QoS, obtained through three different mechanisms: (i) time-scheduling, (ii) priority levels and (iii) bandwidth allocation. Each SpaceFibre node can implement up to 32 VCs and independently configures each one with the desired QoS parameters. Although SpaceFibre packets can be of any length, the data stream is segmented into 64-words frames allowing data interleaving from different virtual channels. When a new frame can be transmitted, SpaceFibre applies its QoS rules to decide which VC is allowed to transmit. Time-scheduling is the first mechanism to be applied. The time is logically split into 64 repeating timeslots of fixed or variable duration. This service guarantees that only the VCs scheduled for the current timeslot can compete for transmission on the shared medium. These VCs are then compared based on their priority level, and only the ones with the highest priority are allowed to be scheduled to transmit. Finally, the bandwidth reservation mechanism is applied. It allows tuning the link utilisation percentage for each VC. The VC with the highest bandwidth credit (i.e. that is using less bandwidth than its assigned portion of the total link bandwidth) is the one that will be scheduled to transmit the next frame.

The SpaceFibre protocol standard has been published by the ECSS [25]. Moreover, it has been incorporated in the latest revision of the ANSI/VITA 78 SpaceVPX backplane standard [30].

References

1. Ademaj, A., Grillinger, P., Steinhammer, K., & Kopetz, H. (2005). The time-triggered ethernet (TTE) design, vol. 2005, pp. 22–33.
2. Anderson, K., Ryan, B., Sonntag, W., Kavvada, A., & Friedl, L. (2017). Earth observation in service of the 2030 agenda for sustainable development. *Geo-Spatial Information Science, 20*(2), 77–96.
3. Armbruster, P., Sheynin, Y., Parkes, S., & Nomachi, M. (2009). SpaceWire missions and architectures. In: *60th International Astronautical Congress 2009, IAC 2009*, pp. 2963–2970. International Astronautical Federation.
4. Bach, H., Mauser, W., & Gernot, K. (2016). Earth observation for food security and sustainable agriculture. *ESASP, 740*, 96.
5. Bandyopadhyay, S., Foust, R., Subramanian, G. P., Chung, S.-J., & Hadaegh, F. Y. (2016). Review of formation flying and constellation missions using nanosatellites. *Journal of Spacecraft and Rockets, 53*(3), 567–578.
6. Bruguier, G., & Palau, J. M. (1996). Single particle-induced latchup. *IEEE Transactions on Nuclear Science, 43*(2), 522–532.
7. Bush, A., Sollmann, R., Wilting, A., Bohmann, K., Cole, B., Balzter, H., Martius, C., Zlinszky, A., Calvignac-Spencer, S., Cobbold, C. A., et al. (2017). Connecting earth observation to high-throughput biodiversity data. *Nature Ecology & Evolution, 1*(7), 0176.
8. Cable, M. L., Hook, S. J., Green, R. O., Ustin, S. L., Mandl, D. J., Lee, C. M., & Middleton, E. M. (2015). An introduction to the NASA hyperspectral infrared imager (hyspIRI) mission and preparatory activities. *Remote Sensing of Environment, 167*, 6–19.

9. Capella Space. Capella cubesat. Available online: https://www.capellaspace.com/.
10. Cappelletti, C., & Robson, D. (2020). Cubesat missions and applications. In: *Cubesat Handbook*, pp. 53–65. Elsevier.
11. Chave, J., Dall, J., Papathanassiou, K., Paillou, P., Rechstein, M., Quegan, S., Saatchi, S., Scipal, K., Shugart, H., Tebaldini, S., Ulander, L., Le Toan, T., & Williams, M. (2018). The biomass mission: Objectives and requirements, vol. 2018-July, pp. 8563–8566.
12. Cozzi, D., Kleibrink, D., Korf, S., Hagemeyer, J., Porrmann, M., Jungewelter, D., & Ilstad, J. (2014). Axi-based SpaceFibre IP core implementation.
13. Crocetti, L., Forkel, M., Fischer, M., Jurečka, F., Grlj, A., Salentinig, A., Trnka, M., Anderson, M., Ng, W.-T., Kokalj, Ž., et al. (2020). Earth observation for agricultural drought monitoring in the Pannonian basin (southeastern Europe): current state and future directions. *Regional Environmental Change, 20*(4), 1–17.
14. Cubesat design specification rev. 12. (2009). Available online: http://www.cubesat.org/resources/.
15. Denis, G., Claverie, A., Pasco, X., Darnis, J.-P., de Maupeou, B., Lafaye, M., & Morel, E. (2017). Towards disruptions in earth observation? new earth observation systems and markets evolution: Possible scenarios and impacts. *Acta Astronautica, 137*, 415–433.
16. Dolce, F., Di Domizio, D., Bruckert, D., Rodríguez, A., & Patrono, A. (2020). Earth observation for security and defense. *Handbook of Space Security: Policies, Applications and Programs*, pp. 705–731.
17. Donlon, C., Berruti, B., Buongiorno, A., Ferreira, M.-H., Féménias, P., Frerick, J., Goryl, P., Klein, U., Laur, H., Mavrocordatos, C., et al. (2012). The global monitoring for environment and security (GMES) sentinel-3 mission. *Remote Sensing of Environment, 120*, 37–57.
18. Drusch, M., Del Bello, U., Carlier, S., Colin, O., Fernandez, V., Gascon, F., Hoersch, B., Isola, C., Laberinti, P., Martimort, P., et al. (2012). Sentinel-2: Esa's optical high-resolution mission for GMES operational services. *Remote sensing of Environment, 120*, 25–36.
19. Elliott, J. R. (2020). Earth observation for the assessment of earthquake hazard, risk and disaster management. *Surveys in Geophysics, 41*(6), 1323–1354.
20. Esposito, M., Conticello, S. S., Pastena, M., & Carnicero Domínguez, B. (2019). In-orbit demonstration of artificial intelligence applied to hyperspectral and thermal sensing from space. In: *CubeSats and SmallSats for Remote Sensing III*, vol. 11131, p. 111310C. International Society for Optics and Photonics.
21. European Cooperation for Space Standardisation. (2008). *Interface and communication protocol for MIL-STD-1553B data bus onboard spacecraft ECSS-E-ST-50-13C*. European Cooperation for Space Standardisation.
22. European Cooperation for Space Standardisation. (2008). *Space product assurance – ASIC and FPGA development, ECSS-E-ST-60-02C*. European Cooperation for Space Standardisation.
23. European Cooperation for Space Standardisation. (2010). *RMAP Standard ECSS-E-ST-50-52C*. European Cooperation for Space Standardisation.
24. European Cooperation for Space Standardisation. (2015). *CANbus extension protocol ECSS-E-ST-50-15C*. European Cooperation for Space Standardisation.
25. European Cooperation for Space Standardisation. (2019). *SpaceFibre – Very high-speed serial link, ECSS-E-ST-50-11C*. European Cooperation for Space Standardisation.
26. European Cooperation for Space Standardisation. (2019). *SpaceWire – Links, nodes, routers and networks, ECSS-E-ST-50-12C Rev.1*. European Cooperation for Space Standardisation.
27. Fanucci, L., Tonarelli, M., Saponara, S., & Petri, E. (2007). Radiation tolerant space wire router for satellite on-board networking. *IEEE Aerospace and Electronic Systems Magazine, 22*(5), 3–12.
28. Fernandez, V., Kirschner, V., Isola, C., Martimort, P., & Meygret, A. (2012). Sentinel-2 multispectral imager (MSI) and calibration/validation, pp. 6999–7002.
29. Ferrell, U., Spitzer, C., & Ferrell, T. (2014). *Digital avionics handbook*. CRC press.
30. Florit, A. F., Villafranca, A. G., McClements, C., Parkes, S., & Srivastava, A. (2017). A prototype SpaceVPX lite (vita 78.1) system using SpaceFibre for data and control planes. In: *2017 IEEE Aerospace Conference*, pp. 1–9. IEEE.

31. Franceschetti, G., & Lanari, R. (1999). *Synthetic aperture radar processing*. CRC Press.
32. Fuller, S., & Cox, T. (2002). Anatomy of a forward-looking open standard. *Computer, 35*(1), 140–141.
33. Gonzalez-Villafranca, A., Ferrer, A., McClements, C., Yu, B., & Parkes, S. (2014). Integrating STAR-Dundee SpaceFibre codec with TI TLK2711.
34. Guanter, L., Kaufmann, H., Segl, K., Foerster, S., Rogass, C., Chabrillat, S., Kuester, T., Hollstein, A., Rossner, G., Chlebek, C., et al. (2015). The EnMAP spaceborne imaging spectroscopy mission for earth observation. *Remote Sensing, 7*(7), 8830–8857.
35. Hemmati, H., Chen, Y., & Some, R. (2013). Multi-Gb/s fiberoptic physical layer for spacecraft interconnects. *Journal of Lightwave Technology, 31*(12), 1899–1905.
36. Huadong, G. (2013). Earth observation in china and the world: History and development in 50 years. *Bulletin of the Chinese Academy of Sciences, 27*(2), 96–98.
37. ICEYE. ICEYE cubesat. Available online: https://www.iceye.com/resources/technology/.
38. IEEE. (2016). IEEE standard for ethernet. *IEEE Std 802.3-2015 (Revision of IEEE Std 802.3-2012)*, pp. 1–4017.
39. Incorporated (ARINC) Aeronautical Radio. Arinc644-part-7. aircraft data network part 7 avionics full duplex switched ethernet (AFDX) network. Available online: http://standards.globalspec.com/std/204622/arinc-664-p7.
40. Kansakar, P., & Hossain, F. (2016). A review of applications of satellite earth observation data for global societal benefit and stewardship of planet earth. *Space Policy, 36*, 46–54.
41. Karatas, Y., & Ince, F. (2016). Fuzzy expert tool for small satellite cost estimation. *IEEE Aerospace and Electronic Systems Magazine, 31*(5), 28–35.
42. Khanvilkar, S., Shah, S. I. A., & Khokhar, A. (2006). RapidIO traffic management and flow arbitration protocol. *IEEE Communications Magazine, 44*(7), 45–52.
43. Kim, Y., Kumar, R., Misra, T., Bhan, R., Rosen, P. A., & Sagi, V. R. (2017). Global persistent SAR sampling with the NASA-ISRO SAR (NISAR) mission. In: *2017 IEEE Radar Conference (RadarConf)*, pp. 0410–0414.
44. Kim, J., Jeong, U., Ahn, M.-H., Kim, J. H., Park, R. J., Lee, H., Song, C. H., Choi, Y.-S., Lee, K.-H., Yoo, J.-M., et al. (2020). New era of air quality monitoring from space: Geostationary environment monitoring spectrometer (gems). *Bulletin of the American Meteorological Society, 101*(1), E1–E22.
45. KP Labs. Intuition cubesat. Available online: https://www.kplabs.pl//.
46. L3Harris. Hypercube cubesat. Available online: https://www.harris.com/sites/default/files/downloads/solutions/55493_hypercube_data_sheet_v2_2_final_.pdf.
47. Lagasio, M., Parodi, A., Pulvirenti, L., Meroni, A. N., Boni, G., Pierdicca, N., Marzano, F. S., Luini, L., Venuti, G., Realini, E., et al. (2019). A synergistic use of a high-resolution numerical weather prediction model and high-resolution earth observation products to improve precipitation forecast. *Remote Sensing, 11*(20), 2387.
48. Lavacca, F. G., Listanti, M., Eramo, V., & Caporossi, S. (2018). Definition and performance evaluation of an advanced avionic TTEthernet architecture for the support of launcher networks. *IEEE Aerospace and Electronic Systems Magazine, 33*(9), 30–43.
49. Li, X. (2015). The first sentinel-1 SAR image of a typhoon. *Acta Oceanologica Sinica, 34*(1), 1–2.
50. Loizzo, R., Guarini, R., Longo, F., Scopa, T., Formaro, R., Facchinetti, C., & Varacalli, G. (2018). Prisma: The Italian hyperspectral mission. In: *IGARSS 2018-2018 IEEE International Geoscience and Remote Sensing Symposium*, pp. 175–178. IEEE.
51. Louet, J., & Bruzzi, S. (1999). Envisat mission and system. In: *IEEE 1999 International Geoscience and Remote Sensing Symposium. IGARSS'99 (Cat. No.99CH36293)*, vol. 3, pp. 1680–1682.
52. Mason, A., & Parkes, S. (2014). Using SpaceWire with LabVIEW.
53. Mathieu, P.-P., & Aubrecht, C. (2018). *Earth observation open science and innovation*. Springer Nature.
54. Matteoli, S., Diani, M., & Corsini, G. (2010). A tutorial overview of anomaly detection in hyperspectral images. *IEEE Aerospace and Electronic Systems Magazine*, 25(7):5–28.

55. McClements, C., McLaren, D., Florit, A. F., Parkes, S., & Villafranca, A. G. (2015). SpaceFibre: A multi-gigabit/s interconnect for spacecraft onboard data handling, vol. 2015-June.
56. McClements, C., McLaren, D., Youssef, B., Ali, M. S., Florit, A. F., Parkes, S., & Villafranca, A. G. (2016). SpaceWire and SpaceFibre on the Microsemi RTG4 FPGA. volume 2016-June.
57. Moreira, A., Prats-Iraola, P., Younis, M., Krieger, G., Hajnsek, I., & Papathanassiou, K. P. (2013). A tutorial on synthetic aperture radar. *IEEE Geoscience and Remote Sensing Magazine, 1*(1), 6–43.
58. Nannipieri, P., Leoni, A., & Fanucci, L. (2019). VHDL design of a spacefibre routing switch. *IEICE Transactions on Fundamentals of Electronics, Communications and Computer Sciences, E102A*(5), 729–731.
59. NASA. Example of SAR image. Available online: https://en.wikipedia.org/wiki/Synthetic-aperture_radar.
60. NASA Nicholas M. Example of SAR image. Available online: https://en.wikipedia.org/wiki/Hyperspectral_imaging.
61. Nguyen, L. H., Joshi, D. R., Clay, D. E., & Henebry, G. M. (2020). Characterizing land cover/land use from multiple years of Landsat and MODIS time series: A novel approach using land surface phenology modeling and random forest classifier. *Remote Sensing of Environment, 238*, 111017.
62. Nieke, J., & Rast, M. (2018). Towards the Copernicus hyperspectral imaging mission for the environment (CHIME). In: *IGARSS 2018-2018 IEEE International Geoscience and Remote Sensing Symposium*, pp. 157–159. IEEE.
63. Normand, E. (1996). Single-event effects in avionics. *IEEE Transactions on Nuclear Science, 43*(2), 461–474.
64. Oliver, C., & Quegan, S. (2004). *Understanding synthetic aperture radar images*. SciTech Publishing.
65. Parkes, S., & Armbruster, P. (2005). SpaceWire: A spacecraft onboard network for real-time communications, vol. 2005, pp. 6–10.
66. Poghosyan, A., & Golkar, A. (2017). Cubesat evolution: Analyzing cubesat capabilities for conducting science missions. *Progress in Aerospace Sciences, 88*, 59–83.
67. Postel, J., et al. (1981). RFC 791: Internet protocol.
68. Postel, J., et al. (1981). RFC 793: Transmission control protocol.
69. Sabater, N., Tenjo, C., Vicent, J., Alonso, L., Rivera, J. P., & Moreno, J. (2014). Synthetic scene simulator for hyperspectral spaceborne passive optical sensors. application to ESA's FLEX/sentinel-3 tandem mission. In: *2014 6th Workshop on Hyperspectral Image and Signal Processing: Evolution in Remote Sensing (WHISPERS)*, pp. 1–4. IEEE.
70. SAE International. Time-triggered ethernet. Available online: https://www.sae.org/standards/content/as6802/.
71. Saifudin, M. A., Karim, A., et al. (2018). Lapan-a4 concept and design for earth observation and maritime monitoring missions. In: *2018 IEEE International Conference on Aerospace Electronics and Remote Sensing Technology (ICARES)*, pp. 1–5. IEEE.
72. Selva, D., & Krejci, D. (2012). A survey and assessment of the capabilities of Cubesats for earth observation. *Acta Astronautica, 74*, 50–68.
73. Shippert, P., et al. (2004). Why use hyperspectral imagery? *Photogrammetric Engineering and Remote Sensing, 70*(4), 377–396.
74. Sinelnikov, V., Kuznetsov, V., & Alpert, S. (2014). Sputnik 1 and the first satellite ionospheric experiment. *cosp, 40*:C0–2.
75. SpaceFab. Spacefibre cable datasheet. Available online: http://www.spacefab.us/space-telescopes.html.
76. Space AVionics Open Interface aRchitecture (SAVOIR). (2019). *SAVOIR On-board Communication System Requirement Document, SAVOIR-GS-008*. Space AVionics Open Interface aRchitecture (SAVOIR).
77. Teodosiu, D., Pistritto, J. C., Soles, L. R., & Boyen, X. (2006). Transmission control protocol, November 28. US Patent 7,143,131.

78. Terraillon, J.-L. (2016). Savoir: Reusing specifications to favour product lines.
79. Thompson, P. T., Corazza, G. E., Vanelli-Coralli, A., Evans, B. G., & Candreva, E. A. (2011). 1945–2010: 65 years of satellite history from early visions to latest missions. *Proceedings of the IEEE, 99*(11), 1840–1857.
80. United Nations. The 17 goals. https://sdgs.un.org/goals.
81. Villela, T., Costa, C. A., Brandão, A. M., Bueno, F. T., & Leonardi, R. (2019). Towards the thousandth cubesat: a statistical overview. *International Journal of Aerospace Engineering*. https://doi.org/10.1155/2019/5063145.
82. World Metrological Organization. Current and future satellites for meteorological and earth observation purposes. Available online: https://www.wmo-sat.info/oscar/satellites.
83. Wye, L., Lee, S., Buonocore, J., Stevens, T., Novosolev, R., Watt, D., Chen, S., Wilhelm, J., Rennich, P., & Zebker, H. (2016). Sri cubesat imaging radar for earth science (SRI-CIRES). In: *Earth Sci. Technol. Forum*.
84. Xu, Y. N., Xu, G. B., Wang, H. B., Chen, L., Bi, J. S., & Liu, M. (2017). Total ionization dose effects on charge-trapping memory with al 2 o 3/hfo 2/al 2 o 3 trilayer structure. *IEEE Transactions on Nuclear Science, 65*(1), 200–205.
85. Yu, P., Koga, R., & George, J. (2008). Single event effects and total dose test results for TI TLK2711 transceiver. In: *2008 IEEE Radiation Effects Data Workshop*, pp. 69–75. IEEE.
86. Yu, P., Koga, R., & George, J. (2008). Single event effects and total dose test results for TI TLK2711 transceiver, pp. 69–75.

Chapter 2
The SpaceFibre Standard

2.1 The OSI Model

The Open Systems Interconnection (OSI) model was standardised by the International Organisation for Standardisation (ISO) in 1984 [1], and it defines a standard protocol stack to facilitate the interoperability of diverse network systems. Figure 2.1 shows the OSI model, including seven abstract layers that completely define the functionality of a communication system, from the transmission of raw data over a physical medium to the highest-level operations. Each intermediate layer is served by the layer below it and implements a set of services that is exploited by the layer above it. The layering approach provides a methodology for grouping similar functions in the same layer, decomposing a complex network system into smaller and more manageable parts. The OSI model is composed of the following layers [1]:

- the *Physical* level defines how to transmit raw data on the physical medium, and it also provides electric and mechanical properties of connectors and cables (e.g. optical fibres, copper cables, wireless systems, etc.).
- the *Data-Link* level is responsible for delivering data to nodes that are at the same level in the network hierarchy, and for framing data to be transmitted. A Data-Link is reliable if it provides mechanisms for ensuring the correctness of received data packets. A reliable Data-Link shall detect errors that can accidentally occur during the transmission process (e.g. a bit-flip) and retransmit corrupted data frames. It shall also implement a flow control mechanism to check if the receiver is ready for receiving new data frames avoiding overflows. The Data-Link layer implements QoS, enabling different classes of traffic to share the same communication link.
- the *Network* layer level forwards variable length data packets from a source to a destination of a network, selecting the shortest possible path. An address uniquely identifies every node.

P. Nannipieri et al., *Next-Generation High-Speed Satellite Interconnect*,
https://doi.org/10.1007/978-3-030-77044-0_2

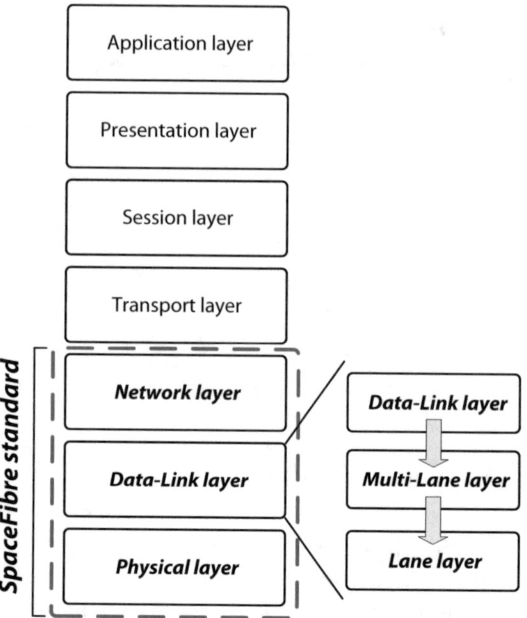

Fig. 2.1 The OSI model

- the *Transport* level provides communication between processes executed on different machines. It can compensate for the lack of services that are not supported by lower layers (e.g. reliability).
- the *Session* level is responsible for opening, closing and managing a session. A session consists of a temporary exchange of information between two user applications. Sessions can be half-duplex, if the traffic is allowed in one direction at a time, or full-duplex, if traffic is allowed in both directions at the same time.
- the *Presentation* level is responsible for formatting data to be presented to the Application level (e.g. ASCII character code translation).
- the *Application* level is the one the user interacts with.

The SpaceFibre (SpFi) standard includes the three lowest levels of the OSI model, implementing the Network layer, the Data-Link layer and the Physical layer. The Data-Link layer is further divided into sub-layers, and it also features the Multi-Lane layer and the Lane layer. In the next sections, the most relevant features of the SpFi protocol stack will be described.

2.2 Physical Layer

The Physical layer handles the reception and the transmission of raw data over a physical medium, also defining connectors, cables and cable assemblies. The SpFi Physical layer is compatible both with electrical and fibre optic cable, and it shall provide two services:

- the Transfer Symbol Level Service, which is responsible, on the transmitter side, for serialising Lane layer 8B/10B symbols [2, 10] and transmitting them over the physical medium (*TX Signal*) and, on the receiver side, for de-serialising incoming data in 10-bit parallel data, and for passing them to the Lane layer (*RX Signal*).
- the Control Service, which is responsible for controlling the operations of the line drivers, line receiver and SerDes, also reporting their status.

The Physical layer shall also recover the clock and data from the received serial bit stream and accept service request from the Lane layer, for enabling line driver, line receiver and line data recovery clock. The Lane layer can also require to invert the polarity of each bit of the output of the de-serialiser, asserting the *RX polarity* signal. The signal *Loss of Signal* is asserted when the RX Signal is below a minimum threshold and it is not strong enough or when the link is too faulty and is not possible to operate reliably. Two signals of the Management Information Base control Physical layer for testing purposes. When the *Near-end serial loopback* signal is asserted, the output bit stream from the serialiser is connected directly into the serial input of the de-serialiser. When the *Far-end serial loopback* signal is asserted, the signal from the output of the line receiver is connected directly to the input of the line driver. Figure 2.2 shows the interfaces to the Physical layer.

The data signalling rate shall be the same for both the directions of a lane, with a ±0.01% tolerance. There is no limitation for the data signalling rate of a SpFi lane,

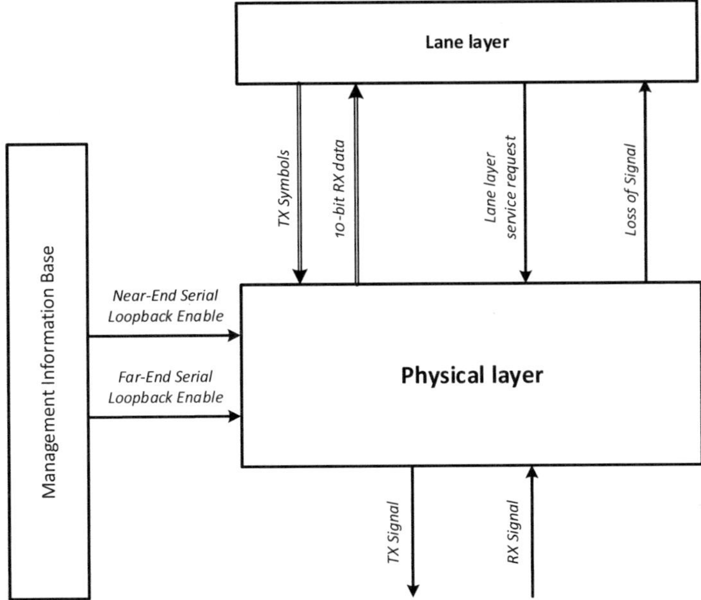

Fig. 2.2 Interfaces to the Physical layer

but the standard suggests adopting one of the following: 1 Gbps, 1.25 Gbps, 2 Gbps, 2.5 Gbps, 3.125 Gbps, 5 Gbps or 6.25 Gbps.

The Physical layer also defines four types of SpFi connectors, cables and cable assemblies for an electrical medium and three for a fibre optic medium [4]. For example, the electrical medium Type-A and Types-B specify parts for flight use and Type-C and Type-D for electronic ground support applications.

2.3 Lane Layer

The Lane layer is responsible for establishing the communication over a SpFi link through a dedicated mechanism that allows synchronising the operations between the far-end and the near-end of the link [4]. The Lane layer is also responsible for encoding data to be transmitted in symbols, and for decoding received symbols in data words with 8B/10B [2]. Furthermore, the Lane layer operates to re-establish the communication between the two ends of a link when an error occurs over a lane. The Lane layer shall provide the following services:

- the *Transfer Service* that provides support for sending and receiving data over a SpFi link.
- the *Control Service* that controls the operations of a lane providing the current state of a lane (receiving only, transmitting only or bi-directional lane), resetting it when necessary.
- the *Capability Service* that sets the capability of the lane at the near-end and reports the capability of the far-end, guaranteeing an efficient communication between the two ends. The capability includes details regarding the nature of the sending lane, indicating if it is part of a Multi-Lane link or a routing switch. Furthermore, the capability informs if incoming data are scrambled for reducing electromagnetic emission or not [8].

Figure 2.3 shows the interfaces to the Lane layer of a single lane SpFi link. In this configuration, the Multi-Lane layer is bypassed, and the Data-Link layer sends and receives a single word at a time. The Lane layer shall accept service requests from the Data-Link layer and can use services of the Physical layer, as indicated by the OSI model [1]. The Data-Link can assert the *Lane Reset* signal for resetting the lane. The *Lane Capabilities* signal is exploited for setting the capabilities of the lane at the near-end and reporting the capability of the far-end. For what concern the interface to the Physical layer, the Lane layer transmits and receives 8B/10B encoded symbols. The *Loss of Signal* indicates that the end has lost signal on its receiver. The *Line Driver Enable* and *Line Receiver Enable* signals activate/de-activate the line driver and line receiver, respectively. The *Clock Data Recovery (CDR) enable* signal can disable the CDR circuitry, in particular, conditions to save power. Finally, when the *Invert RX Polarity* signal is asserted all the bits of the *10-bit RX data* signal are inverted. Figure 2.4 shows the interfaces to the Lane layer of a Multi-Lane SpFi link. The Multi-Lane configuration allows parallelizing the

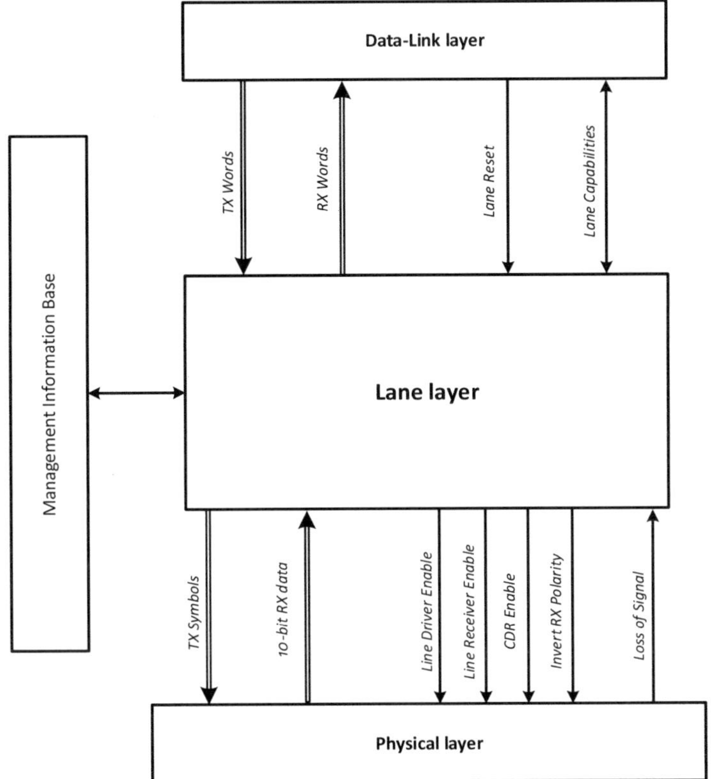

Fig. 2.3 Interfaces to the Lane layer for a single lane link

communication, transmitting *N_LANES* words at a time. The *Lane management* signals allow to implement the *Control Service*, and *LaneState* reports to the Multi-Lane layer the status of each lane. In both the presented configurations, the interface to the Management Information base is exploited for reporting the status of each lane and controlling the initialisation process.

The Lane layer is responsible for establishing the communication across a SpFi link through the Lane Initialisation Finite State Machine (FSM), whose simplified scheme is shown in Fig. 2.5. The *Standby state* is entered at start-up when the Data-Link layer or the Multi-Lane layer asserts the *Lane Reset* signal. The operations of a Lane layer start only when the Management Information base sets the *Lane Start* signal. The initialisation handshake process between the lane layers of the two ends of a SpFi link is obtained through the Lane layer initialisation control words INIT1, INIT2 and INIT3, which identify three different states of the FSM. During the *Started state*, the Lane layer sends one INIT1 control word followed by 64 pseudo-random generated words to the far-end of the link. The Lane initialisation state machine leaves this state only when 1023 valid words and at least one valid INIT1

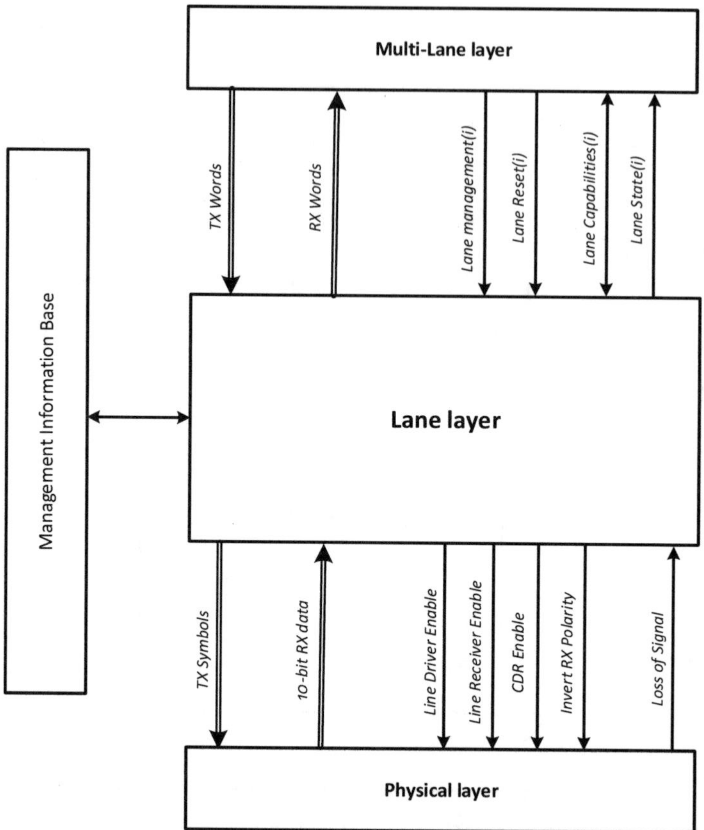

Fig. 2.4 Interfaces to the Lane layer for a multi-lane link

control word or one valid INIT2 control word has been received. In the *Connecting state*, the Lane Initialisation FSM continuously sends INIT2 control words, and it enters the *Connecting state* once three valid INIT2 or INIT3 control words have been received. The INIT3 control word includes lane capability information that shall be passed to the upper layer. The handshake process ends in the *Connecting state*, whit the Lane Initialisation FSM sending INIT3 control words to the far end of the link. The Lane Initialisation FSM enters the *Active state* once three valid INIT3 are received. In the *Active state*, the lane layers of the two ends of the link are properly initialised and synchronised, and the upper layers of the CoDec are finally allowed to send and receive data words. The FSM can leave the *Active state* if the Management Information base de-asserts the *Lane Start* signal or if the Lane layer loses the signal on its receiver, and the *Loss of Signal* is set at "1". For more details about the Lane Initialisation FSM please refer to the SpaceFibre standard, which contains a complete description of it [4]. The Lane layer is also responsible for compensating differences that may arise between the data signalling rate of the

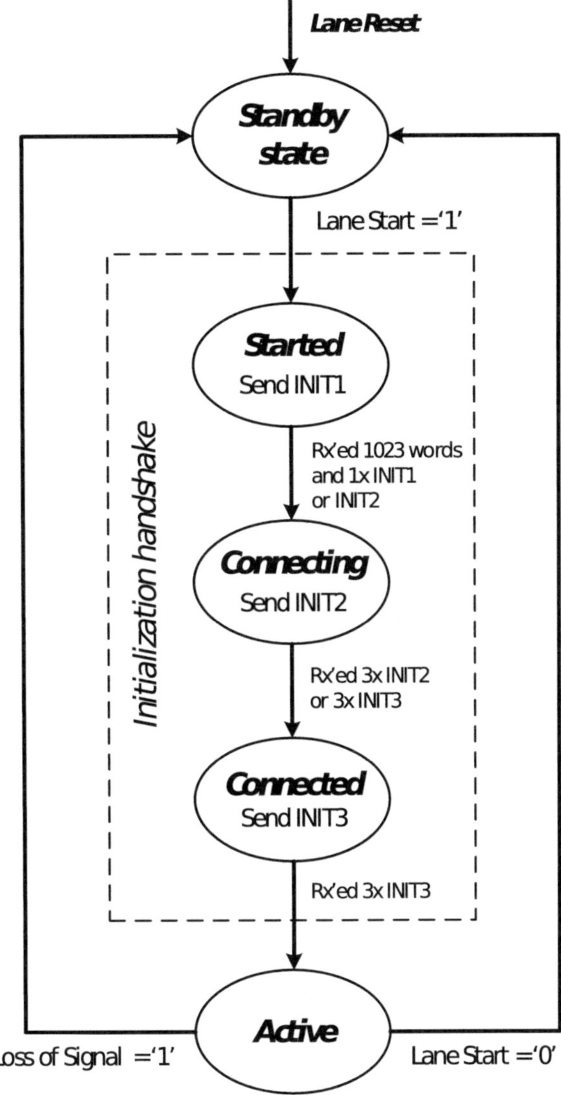

Fig. 2.5 A schematic representation of the Lane Initialisation Finite State Machine

two ends of the communication link. This may happen due to slight differences in the local clock of the two ends of the link. Therefore a receive Buffer is requested by the standard to accommodate small differences between the two clocks. The Elastic Buffer recognises a special SKIP word in the data stream and, if the buffer is more than half-full, the SKIP word is effectively skipped. If, on the other hand, the buffer is less than half-full, the SKIP word is not skipped. One SKIP word is transmitted every 5000 words. This mechanism allows absorbing 100 ppm difference between local oscillators at the two ends. Finally, the Lane layer is responsible for encoding (in transmission) and decoding (in reception) data and control words into symbols through the 8B/10B [2] coding/encoding. This technique maps 8-bit words to 10-bit symbols to achieve DC-balance and bounded disparity [10], allowing also to detect soft error such as bit-flips and implementing part of the SpFi FDIR system.

2.4 Multi-Lane Layer

2.4.1 Multi-Lane Link Basic Concepts

A Multi-Lane link is composed of several lanes running in parallel. Figure 2.6 shows the example of a typical Multi-Lane link, as presented in [7]. We define the following parameters for a better understanding of Multi-Lane basic concepts:

- a *port* is a Multi-Lane endpoint with input and output interfaces.
- a *lane* is a bi-directional physical connection allowing the exchange of information between two ports.
- a *link* incorporates one or more lanes.
- a *hot redundant lane* is a lane in the idle state that does not transmit useful data in normal conditions. It is promoted to the role of data-sending lanes only in case of a lane failure. A hot redundant lane continues to be active as long as the failing lane remains inactive.
- a *symbol* is a set of an arbitrary number of bits (protocol-dependent), which will represent a numerical value handled by the protocol atomically.
- a *word* is a set of an arbitrary number of *symbols*.
- a *row* is an aggregation of one or more words. The number of words composing a row corresponds to the number of lanes transmitting data at a given time.

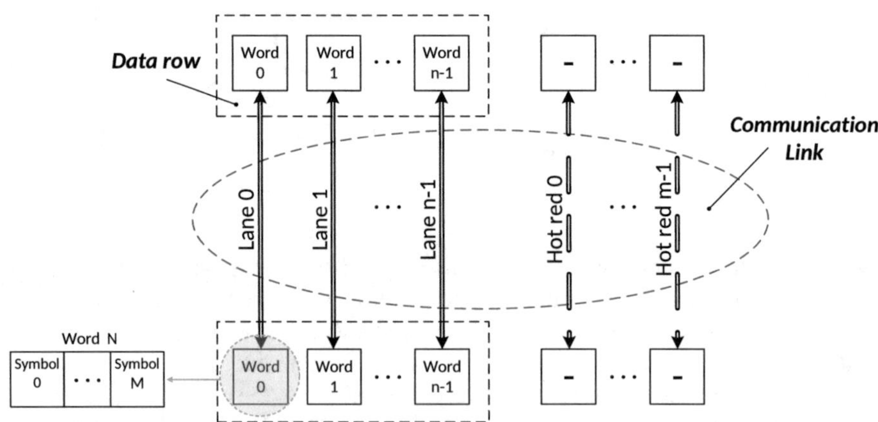

Fig. 2.6 An example of Multi-Lane communication link. A data row is transmitted over the active lanes. Hot redundant lanes are promoted to the role of the data-sending lane in case of lane failures [7]

In the example of Fig. 2.6, the communication link is composed of n regular lanes and m hot redundant lanes for a total of $N\_LANES = n + m$. Each lane of the Multi-Lane link transmits a data word belonging to a data row. Slight delays may be present between lanes because they are separated at the physical level. Thus, the reception circuitry shall guarantee a coherent data reception, providing a data alignment system able to pass to the upper layers of the protocol data belonging to the same row, ensuring data consistency. In case of a lane failure, the bandwidth of the communication link is not necessarily degraded since a hot redundant lane can replace the failing lane. Furthermore, a hot redundant lane can be initialized as a normal lane, guaranteeing a low latency during the replacement process.

2.4.2 SpaceFibre Multi-Lane Layer

The Multi-Lane layer is responsible for supporting the communication over a link composed of up to 16 lanes. The Multi-Lane layer allows drastically increasing the bandwidth of a SpFi link, which is proportional to the number of lanes. It also enhances system robustness and reliability since in case of one or more lane failure the traffic shall be re-directed over the remaining active lanes. The Multi-Lane layer provides the following services:

- the *Link reset Service* that resets all the lanes of the link.
- the *Capability Service* that sets the capability of the lane at the near-end and reports the capability of the far-end.
- the *Transfer Service* that provides support for sending and receiving data over a SpFi link.

The Multi-Lane layer shall accept services request from the Data-Link layer and can use services of the Lane layer. The interfaces to the Multi-Lane layer are shown in Fig. 2.7. When the *Link Reset* signal is asserted, the Multi-Lane layer shall reset all the lanes of the link. The *Lane Capabilities* signal is used for setting the capabilities of the near-end and reporting the capabilities of the far-end of the link. The Management Information base interface is exploited for reporting the alignment state of the link, and the operating mode of each lane (transmission only, receiving only or bi-directional lane), which is established by the Multi-Lane layer itself. The Multi-Lane layer is optional, and it can be bypassed in the case of a single lane link implementation.

More specifically, the Multi-Lane layer is responsible for:

- Verifying lanes alignment. When data communication starts, the Multi-Lane layer shall guarantee that data words received from lane layers are properly aligned.
- Receiving and transmitting a data row composed of n data word from/to the Data-Link layer.
- Receiving and transmitting a single data word from/to n lane layers.

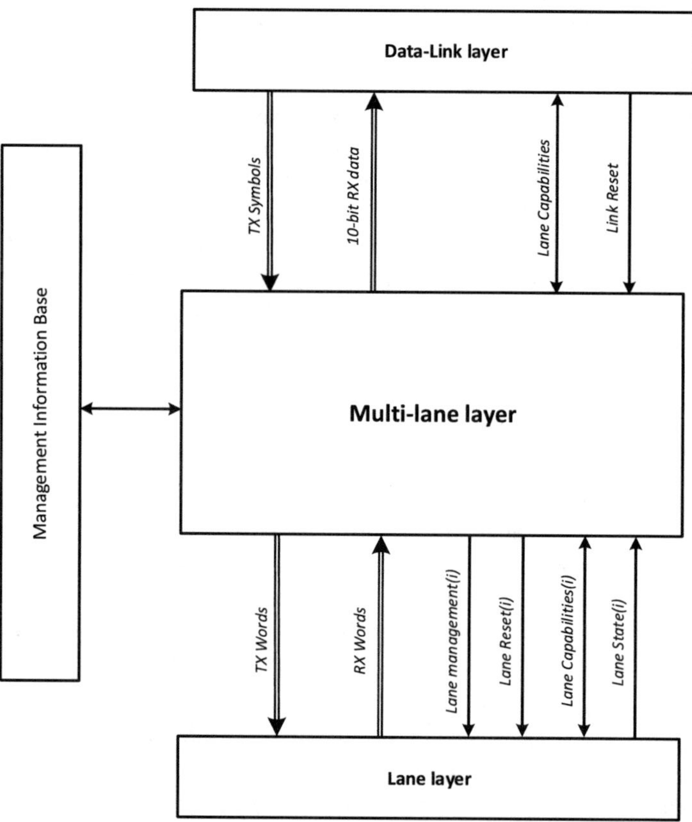

Fig. 2.7 Interfaces to the Multi-Lane layer

- Informing the Data-Link layer about what and how many lanes are active at a given time.
- Managing optional *mono-directional* and *hot redundant* lanes.

As indicated in [7], data alignment is necessary to preserve data coherency. When data communication starts, the Multi-Lane layer controls if data words are perfectly aligned over all the active transmitting lanes. Indeed, small delays (2–3 clock cycles at most) can be present between lanes, and data words could end up in the wrong row. For this reason, a FIFO buffer (3–4 positions [4]) completes the alignment process exploiting special Multi-Lane control words (ACTIVE and ALIGN). In particular, the Multi-Lane FSM (ML-FSM) is responsible for handling the alignment process, and its scheme is shown in Fig. 2.8. The *Not-Ready state* is entered when the Data-Link layer sets the *Link Reset*. In this state, the alignment process starts, and the ML-FSM repetitively sends 7 ACTIVE control words followed by one ALIGN control word. When a particular bit in the ACTIVE control words is asserted, it indicates that the corresponding lane at the far-end of the link

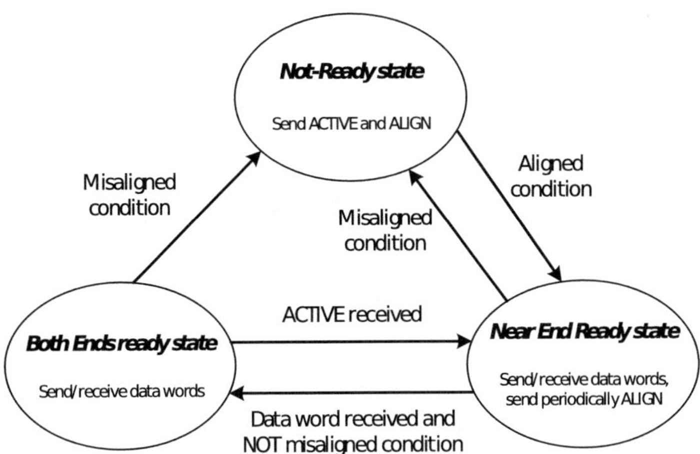

Fig. 2.8 A schematic representation of the Multi-Lane Finite State Machine

is in the *Active state*. The ALIGN control word contains two fields that indicate the identification number of the lane sending the ALIGN control word and the number of lanes that are in the *Active state* at the far-end. The ACTIVE and ALIGN control words are sent simultaneously over all active lanes. The ML-FSM moves to the *Near-End Ready state* when all lanes are aligned, having an ALIGN control word as output, and the number of active lanes at the far-end lanes and near-end is the same. In the *Near-End Ready state*, the ML-FSM sends seven data words interleaved with one ALIGN control word. If at least one valid data word per lane is received and no misaligned conditions occur (e.g. invalid data received, changing in the number of active lanes, etc.) the ML-FMS enters the *Both Ends Ready state* and the Multi-Lane alignment process is finally completed. Instead, if a misaligned condition is detected, the ML-FSM returns to the *Not-Ready state*. In the *Both Ends Ready state*, the Multi-Lane layer pass data words to/from the Data-Link layer without sending Multi-Lane control words, and the CoDec can safely transmit and receive data over a SpFi link. The ML-FSM moves from the *Both Ends Ready state* to the *Not-Ready state* if a misaligned condition is detected, and to the *Near-End Ready state* if an ACTIVE control word is received, meaning that the far-end alignment process is not completed yet.

Figure 2.9 provides an example of lane alignment. In this use case, the SpaceFibre link is composed of 4 lanes, and a 3-position FIFO buffer is implemented for each lane. As shown in Fig. 2.9a, lane 1 is not aligned with the other lanes of the link, and it is necessary to realign it to receive correct data rows. To realign the link, the ML-FSM de-asserts the read enable (re) of the FIFO buffer in which an ALIGN control word was previously read, as shown in Fig. 2.9b. After one clock cycle also lane 1 provides an ALIGN control word, the link is finally aligned, and the Multi-Lane layer is now able to receive coherent data rows (Fig. 2.9c).

SpFi Multi-Lane shall support a generic number of lane failures. The only mandatory condition to preserve the communication over the link is that at least

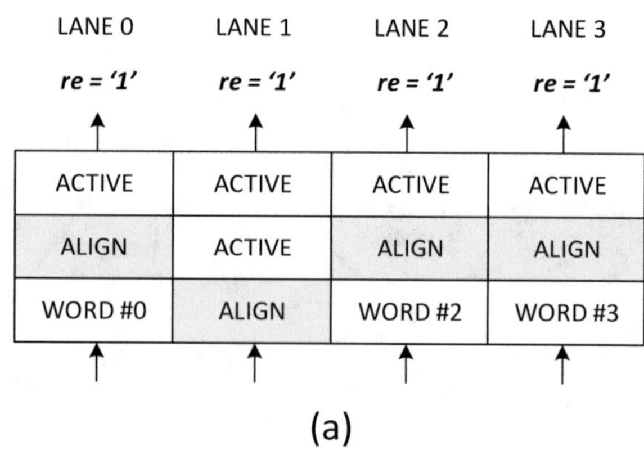

LANE 0	LANE 1	LANE 2	LANE 3
re = '0'	*re* = '1'	*re* = '0'	*re* = '0'
ALIGN	ACTIVE	ALIGN	ALIGN
WORD #0	ALIGN	WORD #2	WORD #3
WORD #4	WORD #1	WORD #6	WORD #7

(b)

LANE 0	LANE 1	LANE 2	LANE 3
re = '1'	*re* = '1'	*re* = '1'	*re* = '1'
ALIGN	ALIGN	ALIGN	ALIGN
WORD #0	WORD #1	WORD #2	WORD #3
WORD #4	WORD #5	WORD #6	WORD #7

(c)

Fig. 2.9 An example of Multi-Lane alignment process

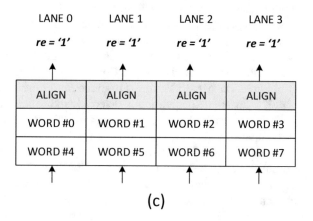

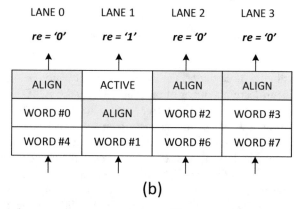

one bi-directional lane shall be active at any time. Thus, the Multi-Lane layer shall continuously monitor the status of each lane and refer to the Data-Link layer what and how many lanes are active to guarantee consistent data transmission and reception.

The SpFi standard supports *mono-directional* lanes that differ from bi-directional lanes because they are allowed to either transmit or receive data. Mono-directional lanes may find application in asymmetric links, where the bandwidth required in one direction is much higher than in the other (e.g. links serving high-bandwidth payload), allowing to save power consumption. Moreover, SpFi supports for *hot redundant* lanes. STAR-Dundee company holds a patent [5] in the SpaceFibre multi-lane technology. However, a licence is guaranteed for free within the space domain.

2.4.3 High-Speed Protocol for Space and Multi-Lane Features

As previously introduced, Multi-Lane is a powerful tool for enhancing system robustness and reliability, also increasing the bandwidth of the communication system. The following section will provide a brief analysis of the Multi-Lane features supported by the newest high-speed protocols introduced in Sect. 1.3, focusing on RapidIO and SpaceFibre (TTEthernet does not support multi-laning). RapidIO handles up to 16 bi-directional lanes, with a maximum data rate of 6.25 Gbps per lane, providing 1x, 2x, 4x, 8x and 16x lane configurations [9]. A RapidIO link shall support the 1x configuration and may support 2x, 4x, 8x and 16x modes. 1x mode is mandatory to guarantee that two ports with different width can share a common link width that they can use to communicate with each other. Links including two or more lanes shall have a redundancy lane to provide a fallback in case of a lane failure. The redundancy lane is lane number 1 for the 2x mode and lane number 2 for the other configurations. This feature works only for pairs of ports that support the same redundancy lane (e.g. it does not work for a 2x port connected to a 4x port). Mono-directional lanes can be instantiated for supporting asymmetric links.

SpaceFibre supports up to 16 bi-directional lanes, with an unbounded maximum data rate (with current transceivers in the order of 6.25 Gbps per lane). A communication link includes *N_LANES* lanes with $1 \leq N\_LANES \leq 16$, and the two ends of the link shall be composed of the same number of lanes. In case of a lane failure, the communication process shall continue on the remaining active lanes. The protocol provides a mechanism to handle an arbitrary number of lane breakdowns, and the communication process shall continue until one bi-directional lane is active. SpFi supports asymmetric links and hot redundant lanes, as previously explained.

RapidIO supports communication links with ports featuring a different number of lanes, making the integration of a RapidIO port in a communication link straightforward. Indeed, two RapidIO ports shall always communicate exploiting the 1x configuration. On the contrary, the ports of a SpFi link shall have the same number of lanes. However, the SpFi Multi-Lane apparatus appears to be more flexible than RapidIO because it supports designs with an arbitrary number of lanes between 1 and 16, allowing to tune data rate with high precision with consequent benefits in terms of hardware resources needed for system realisation and power consumption. Furthermore, it offers a graceful degradation of bandwidth performance in case of a lane failure since only the failing lanes stop to transmit and receive data. RapidIO provides a mechanism to recover from a lane failure, but it supports only a sub-set of SpFi configurations (16x, 8x, 4x, 2x and 1x). Thus, in case of a lane breakdown, RapidIO does not foresee a graceful degradation of the bandwidth that halves every time a lane fails because a RapidIO link shall necessarily work in a valid configuration. For example, if a link is configured in the 16x mode, after a lane failure it will start to work in the 8x mode. Moreover, SpFi supports hot redundant lanes, which are not foreseen by the RapidIO standard. If hot redundant lanes are included, it is possible to avoid any reduction of the available data rate since a hot redundant lane physically substitutes the failing lane in the link. The usage of hot redundant lanes allows building a redundant system, with a limited impact on the hardware resources needed for the implementation of CoDec. Indeed, a hot redundant lane only requires an additional lane layer, which has a limited effect on the overall resources usage [3] and a transceiver. The inclusion of a hot redundant lane does not require to deeply modify the Data-Link layer since the width of the data path is unchanged, and there is no need to reshape the data stream, as for regular Multi-Lane communication. Therefore, the control logic behind the mechanism is straightforward and not heavy to be implemented.

2.5 Data-Link Layer

The Data-Link layer is the most complex layer of the SpFi standard. It implements QoS managing up to 32 VCs carrying independent flows of information over the same physical link. The Data-Link layer is also responsible for most of the SpFi FDIR, detecting accidental error occurred during the transmission process. Indeed, it includes CRC, an error-detecting code [6], which is used to detect accidental changes to raw data that are common in the space environment due to radiations. In case a corrupted data frame is detected, an error is signalled to the far-end of the link that provides for the retransmission of that data frame. Moreover, the Data-Link layer is also responsible for embedding data to be transmitted in SpFi data frames. Optional data scrambling may be included in the Data-Link layer for reducing electromagnetic emissions [8].

The Data-Link layer provides the following services:

- the *Virtual Channel Service* for supporting up to 32 independent VCs.
- the *Broadcast Service* for sending and receiving BroadCast (BC) messages over a SpFi network. Broadcast messages are short messages that are received by all the nodes of a SpFi network for synchronisation purposes.
- a *Schedule Synchronisation Service* for synchronising the scheduling of VCs.

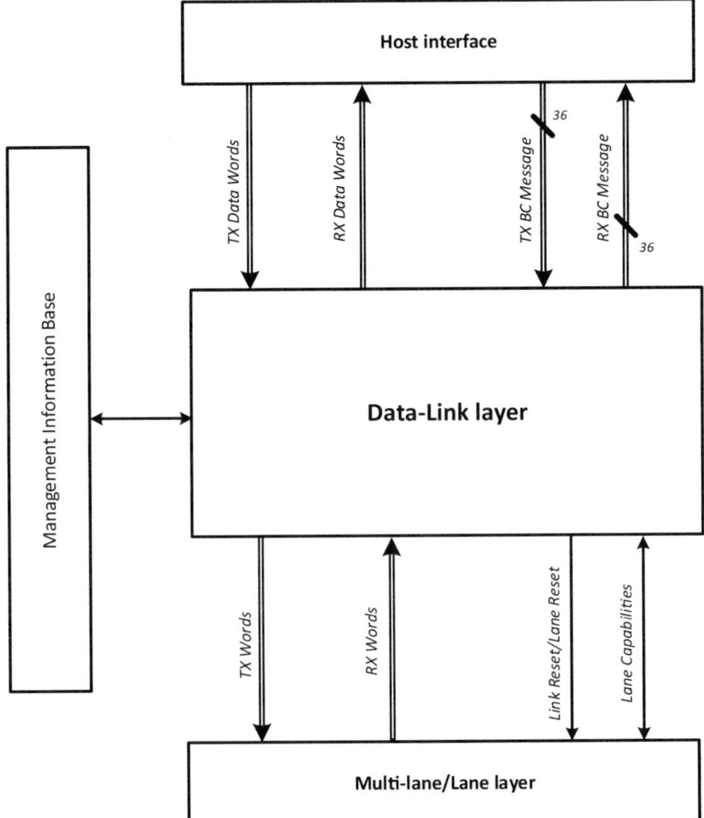

Fig. 2.10 Interfaces to the Data-Link layer

The Data-Link layer shall accept service requests from the Network layer (not included in the proposed implementation) and can use Lane layer (single lane configuration) or Multi-Lane (two or more lanes) services.

Figure 2.10 shows the interfaces to the Data-Link layer. The interface to the host includes the signals *TX Data Words* and *RX data Words*, which are related to VCs data traffic, handling N_LANES words in parallel, with $N\_LANES = 1$ for the single lane configuration and $N\_LANES \geq 2$ for the multi-Lane configuration. The signals *TX BC Message* and *RX BC Message* implement the broadcast interface to the host interface. The Management Information base interface can set the QoS parameter for each VC and data scrambling, and receive status information regarding the VCs and error information. The interfaces to the Multi-Lane and Lane layer have been described in previous sections. The inclusion of the Multi-Lane layer in the protocol stack determines major adjustments in the architecture of the Data-Link layer, to adapt it to support Multi-Lane layer features. In particular, VC interfaces shall be enlarged to provide a complete data row per clock cycle.

Similarly, the CRC and data scramble calculation shall be performed over multiple words at a time. Moreover, the Data-Link layer requires a mechanism for handling data flow in case of one or more lane failures. Let us call the number of inactive lanes at a given time n_{fail}. In that case, the link could transmit only $n - n_{fail}$ words per clock cycle. However, the VC interface continues to provide n words at a time, and the Data-Link layer shall manage reading operation and synchronise operations to avoid the transmission of inconsistent data or even data loss. The Data-Link layer shall handle a generic number of lane failures, guaranteeing a coherent data transmission.

2.5.1 SpaceFibre Quality of Service

The expression QoS refers to a series of mechanisms that allows different classes of data traffic to coexist on the same link. The SpFi standard includes priority, bandwidth reservation and scheduled QoS:

- *Priority*: SpFi defines several levels of priority, and a VC is allowed to send data only when no VC with higher priority is ready for transmitting. If two or more VCs have the same priority, the MAC schedules the VC with the highest bandwidth credit (see Bandwidth reservation). If the SpFi link is set for operating only with priority, each VC shall be assigned a different priority level. Best effort service can be obtained on the VC with the lowest priority level.
- *Bandwidth reservation*: one of the main concerns of a system engineer designing the communication infrastructure of a spacecraft is knowing the available bandwidth for each communication link; e.g. if the payload produces a certain amount of data, the communication protocol shall be able to deal with the necessary communication bandwidth to avoid losing any data. This is something that is not achievable with the priority mechanism, which may lead lower priority endpoint to starvation if a certain amount of bandwidth is not reserved to them. The idea behind the bandwidth reservation is to calculate the bandwidth used by a certain VC and compares it to its reserved bandwidth, to calculate the precedence for that specific VC. This means that if a VC did not use a lot of its reserved bandwidth recently, it will gain higher precedence, to be scheduled. Vice versa, in case a VC uses more than its available bandwidth, its precedence will be lowered. The bandwidth reservation establishes precedence basing on the link bandwidth reserved for each VC (Expected_BW) and its recent link utilisation, which is tracked using the bandwidth credit parameter (BW_credit) as indicated in (2.1):

$$BW\_credit(i) = \sum_{n} Available\_BW(n) - \frac{Used\_BW(n)}{Expected\_BW(i)} \tag{2.1}$$

  where Available_BW(n) is the overall bandwidth for transmitting n data frame or words, Used_BW(n) is bandwidth exploited for transmitting n data frame or words and Expected_BW(i) is the expected bandwidth of VC(i). A bandwidth reserved VC is allowed to send data when no VC with higher priority is ready, and when it has the highest bandwidth credit within its priority level.

- *Scheduled QoS*: to set-up a deterministic Data-Link, SpaceFibre ensures that a VC can be scheduled for communication at a particular time. Time can be split into 64 time-slots, whose duration is in the range from 100 μs to 16 ms, depending on the target application. A VC can be configured to be scheduled in one or more timeslot, depending on its requirement. When a time-slot starts, one or more VCs are allowed to send a data frame, and the MAC selects the one with the highest precedence. Precedence is computed for each VC separately, and it is defined as the sum of VC priority and bandwidth credit. This mechanism enforces the possibility of deterministic communication: the maximum number of VC that can be scheduled within a timeslot is fixed a priori.

Figure 2.11 show a scheme of SpFi QoS. For more details about the SpFi QoS please refer the SpFi standard [4].

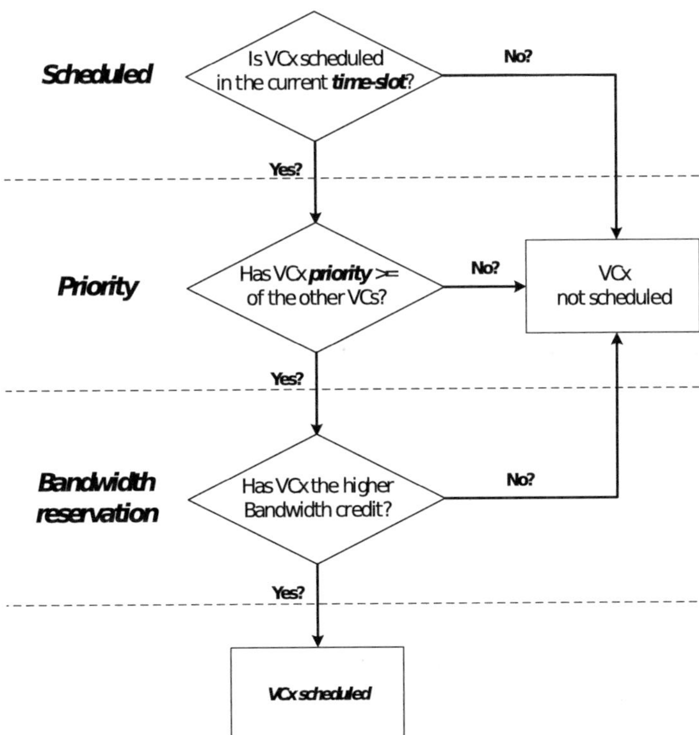

Fig. 2.11 A scheme of the SpFi QoS

2.5.1.1 Bandwidth Credit Example

To help the reader understand the mechanism underneath the bandwidth reservation, in the following we show an example of how the value is calculated and which impact has on the VC scheduling. Let us assume to have a VC with 40% of the overall bandwidth allocated. Consequently, the value of the expected bandwidth will be 0.4. Let us assume that each frame sent is build up by the maximum number of words, which is 64, each made of 4 symbols. The number of symbols sent per frame will be then 256. Let us consider the last 20 frames sent on the link by all the VCs and let us assume that only 1 out of those frames belongs to the VC we are considering. To correctly compute the value of the bandwidth credit, let us compute each parameter of Eq. (2.1):

$$\sum_{n} Available\_BW(n) = 20 \cdot 256 = 5120[Symbols] \tag{2.2}$$

The available bandwidth in the period we are considering is calculated as the number of symbols in a frame, 256, times the number of frames sent, 20.

$$\sum_{n} Used\_BW(n) = 256 \cdot 1 = 256[Symbols] \tag{2.3}$$

The used bandwidth in the last 20 frames is equal to 1 frame or 256 Symbols, as specified: this means that only one frame has been sent from the VC we are taking into account over the last 20 frames.

$$Expected\_BW(i) = 0.4 \tag{2.4}$$

The expected bandwidth is 40%, as previously indicated. With all these values, we are able to compute the Bandwidth credit for the VC we are taking into account. According to Eq. (2.1):

$$BW\_credit = 5120 - \frac{256}{0.4} = 4480 \tag{2.5}$$

Since the VC used such a small fraction of the bandwidth (5%) in the last 20 frames, even if its expected bandwidth is 40%, the bandwidth credit value is very high. Please note that this value is constantly computed for each VC, starting from the very first data frame transmission. On top of that, it is important to remember that the bandwidth credit shall saturate at a specific value, called *Bandwidth Credit Limit*, both positively and negatively. If the positive *Bandwidth Credit Limit* is reached, it means that the VC is taking less bandwidth than expected (VC Underused condition), if the negative *Bandwidth Credit Limit* is reached, it means that the VC is taking more bandwidth than expected (VC Overused condition). Both cases indicate a non-nominal behaviour, due to erroneous settings of QoS by the user, and shall be signalled to take actions.

Priority and Time Scheduling Mechanisms are much easier to handle: within a single time frame, the VC scheduled to communicate is chosen following the flux diagram of Fig. 2.11.

2.5.2 *Fault Detection Isolation and Recovery*

We have already mentioned that SpaceFibre is a fault-tolerant communication technology: it means that in case an error occurs, it can re-send data to avoid any data loss. The first level of *Fault Detection* is provided by 8B/10B codes, as specified in the lane layer. This is done with an ACKnowledge (ACK) Not-ACKnowledge (NACK) mechanism: each time that the far-end of the link correctly receives a data packet, a broadcast or a control word, it sends an ACK message, with an identifier named sequence number (SEQ_NUM) uniquely bounded to the specific message; on the other hand, if the received packet/data word is corrupted, it sends a NACK. The near-end shall save in an error recovery buffer all the data not already acknowledged, which are then scheduled to be resent in case a NACK is received. SEQ_NUM is used to understand which packet or control word has been correctly received and which not, being de-facto the indexing mechanism of the recovery buffer. VC buffers shall be scheduled to send data by a Medium Access Controller (MAC). The scheduling done by the MAC is the vital tasks providing QoS to the communication link; data is scheduled to be sent based on priority level and bandwidth allocation.

2.6 Network Layer

The Network Layer describes how to create networks by connecting end-nodes and routers.

The network layer is responsible for packet and broadcast messages transfer over the network, from the Data-Link layer interface of a port to the Data-Link interface on another port. This layer is optional, and it is not used at all in point-to-point communication links. SpaceFibre is a flexible protocol and it supports several network-level features, such as:

- *Multicast*: a packet can be forwarded to multiple output ports at the same time according to its header.
- *Group Adaptive Routing*: the output port of a packet can be chosen from a set of possible output ports, choosing the first available one.
- *VC Timeout*: whenever the transfer of data from an input port to the connected output port requires more time than expected (higher than this timeout), the wormhole is closed, and the remaining part of the packet is discarded.

The SpaceFibre network layer comes with the concept of Virtual Network (VN): a SpaceFibre network can be seen as a set of independent networks called VN, with similar proprieties to SpaceWire networks. Each VN is parallel and independent to the others and it comprises a VC across each link which is part of the VN. Separate VN belonging to the same network infrastructure shares the physical communication medium, i.e. the SpaceFibre network. The VN operates on top of the Data-Link layer, therefore QoS and FDIR mechanism belonging to the Data-Link layer are still present. However, depending on how the Network layer is configured, VN can guarantee a higher level QoS and FDIR mechanism. The concept of VN can be seen as an extension and evolution of the concept of VCs. Each tuple <VC, Port> is associated with a VN and a packet can flow only through VCs belonging to the same VN. An easy way to look at it is to think that each VN is an isolated SpaceWire-like network. A design example of a SpaceFibre network layer based on the VN mechanism can be found later in the text, in Sect. 5.2.2.

Routing Switches are the key elements in a SpaceFibre network: they connect multiple endpoints, being responsible for packets forwarding. The Network layer introduces several concepts characterising a SpaceFibre Network. The packet length is not constrained; it means that packets can be of any length. This requires an appropriate protocol instrument to address problems that may arise if we go the extremes: a series of very small packets strongly increases the protocol overhead, while very long packets may generate starvation problems that demand a timeout mechanism to guarantee regular system operation. A SpaceFibre packet shall comprise one or more data characters followed by an End of Packet (EOP) marker, or Error End of Packet (EEP) marker. The format is identical to the one of SpaceWire. It is shown in Fig. 2.12.

There are two possible addressing schemes: path addressing and logical addressing. It is up to the source node to decide which one to use, and both of them can be used by the same network, depending on its configuration. When using path addressing the source node writes in the header of the packet the output ports index of each Routing Switch along the path to the destination. The drawbacks are that the source node must have complete knowledge of the topology of the network and that in the case of a complex network the length of the address path may introduce significant protocol overhead. On the other hand, the Routing Switches immediately know where to forward a packet with very small logical effort. When using the logical addressing the source node writes only the logical address in the form of an identifier of the destination node. It is up to the Routing Switches to associate this identifier to a certain output port, using their Routing Tables; the forwarding mechanism is the Wormhole Routing: it is a forwarding scheme that

Fig. 2.12 SpaceFibre packet format [4]

when an input port of the router has data to transmit, it connects it to the requested output port if the latter is free. The output port is then no more available for other input ports until the entire packet has traversed it. Therefore, a packet can span along the network keeping multiple ports busy. This policy allows low hardware resource implementations and requires small or no buffers in the routers. On the other hand, it introduces a source of non-determinism, given that a certain output port could be kept busy for an unknowingly long amount of time. This opens a big challenge about the definition of a network-wide Quality of Service and the latency-bounded data delivery, which is currently not fully addressed.

2.7 Management Layer

The management layer (or Management Information Base, MBI) is responsible for the control and configuration of each layer.

Three different interfaces are provided to the user application:

- A *Packet Level Interface* used to send and receive data packets.
- A *Management Interface* used to configure and control the whole SpFi interface.
- A *Broadcast Message Interface* used to broadcast short messages over the network.

The RMAP in compliance with the ECSS-E-ST-50-52 shall be used for remote configuration, control and monitoring of SpaceFibre networks. The standard provides an extensive list of configuration and status parameter which can be set and/or read through these interfaces.

2.8 SpaceFibre Data and Control Words

The D/K notation is used in the SpaceFibre standard to identify characters that are part of a data word or control word. A data word is represented in the format D.xx/y, where xx is the decimal representation of the 5 least significant bits and y of the three most significant bits of the 8-bit character. The most significant bit D is equal to "0". Similarly, a control word is represented in the format K.xx/y, but in this case, K is equal to "1". The D/K notation is shown in Fig. 2.13a.

In particular, each SpaceFibre data word is composed of 4 D.xx/y values, and a SpaceFibre control word has a K.xx/y in the most significant position, followed by three D.xx/y values, as shown in Fig. 2.13b. Tables 2.1, 2.2 and 2.3 show Lane layer, Multi-Lane layer and Data-Link layer control words, respectively.

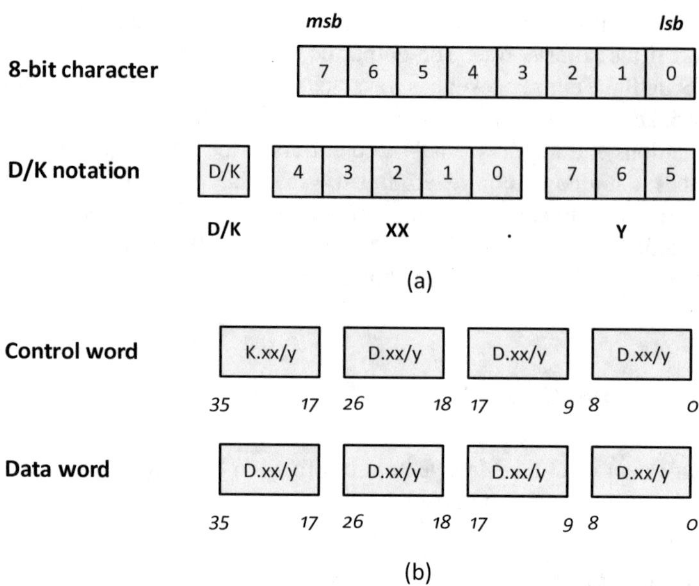

Fig. 2.13 In (**a**) and example of D/K notation. In (**b**) the SpaceFibre format for data and control words

Table 2.1 Lane layer control words

Name	Control word	Function
SKIP	Comma, LLCW, SKIP, SKIP K28.7, D14.6, D31.3, D31.3	Sent every 5000 words for synchronising the elastic buffer operations. The LLCW fields indicates that SKIP is a Lane Layer Control Word
IDLE	Init Comma, LLCW, INIT1, INIT1 K28.5, D14.6, D6.2, D6.2	Sent when neither the Data-Link layer nor the Multi-Layer provides valid words
INIT1	Init Comma, LLCW, INIT1, INIT1 K28.5, D14.6, D6.2, D6.2	Sent as part of the lane synchronisation handshake. The LLCW fields indicates that INIT1 is a Lane Layer Control Word
Inverse INIT1	Init Comma, iLLCW, iINIT1, iINIT1 K28.5, D17.1, D25.5, D25.5	Received as part of the lane synchronisation handshake if the Physical layer signal *Invert RX polarity* is asserted. The iLLCW fields indicates that iINIT1 is an inverse Lane Layer Control Word
INIT2	Init Comma, LLCW, INIT2, INIT2 K28.5, D14.6, D6.5, D6.5	Sent as part of the lane synchronisation handshake after INIT1. The LLCW fields indicates that INIT2 is a Lane Layer Control Word
Inverse INIT2	Init Comma, iLLCW, iINIT2, iINIT2 K28.5, D17.1, D25.2, D25.2	Received as part of the lane synchronisation handshake after iINIT2 if the Physical layer signal *Invert RX polarity* is asserted. The iLLCW fields indicates that iINIT2 is an inverse Lane Layer Control Word

(continued)

Table 2.1 (continued)

Name	Control word	Function
INIT3	Init Comma, LLCW, INIT3, Capability K28.5, D14.6, D24.1, D0.0-D31.7	Sent as part of the lane synchronisation handshake after INIT2. The capability field informs the far-end of the lane of the capability of the near-end node. The LLCW fields indicates that INIT3 is a Lane Layer Control Word
STANDBY	Comma, LLCW, STBY, Reason K28.7, D14.6, D30.3, D0.0-D31.7	Sent when the transmitter is going to tri-state its driver. This mechanism can be exploited to save power when the host has no data to transmit. The Reason field indicates the reason why the STANDY control word was sent. The LLCW fields indicates that STANDBY is a Lane Layer Control Word
LOST_SIGNAL	Comma, LLCW, LOS, Reason K28.7, D14.6, D4.3, D0.0-D2.0	Sent when a node has no signal on its receiver. The Reason field is used to specify the reason why the LOST_SIGNA control word was sent. The LOST_SIGNAL fields indicates that STANDBY is a Lane Layer Control Word
RXERR	Error, Reserved, Reserved, Reserved K0.0, D0.0, D0.0, D0.0	It indicates that the decoder detected an error in the received data stream. Any received word containing one or more symbols unknown symbol is replaced by an RXERR control word

Table 2.2 Multi-lane layer control words

Name	Control word	Function
ACTIVE	Comma, ACTIVE, ACT_LS, ACT_MS K28.7, D0.1, D0.0-D31.7, D0.0-D31.7	Sent as part of the Multi-lane alignment process. The 16 bits of the ACT_LS and ACT_MS fields indicate what lanes are in the active state
ALIGN	Comma, ALIGN, LANES, iLANES K28.7, D23.3, D0.0-D31.7, D0.0-D31.7	Sent as part of during the Multi-lane alignment process. The field LANES indicates the number of lanes of the sending node that are in the active state. The iLANES field is the bit-wise inverse of the LANES field
PAD	Comma, Fill, Fill, Fill K28.7, K27.7, K27.7, K27.7	Sent when the number of words to be sent is lesser than the number of the active lanes in order to form a complete data row

Table 2.3 Data-Link layer control words

Name	Control word	Function
SDF	Comma, SDF, VC, Reserved K28.7, D16.2, D0.0-D31.0, D0.0	Start of data frame control word. The VC field contains the number of the VC sending the data frame
EDF	EDF, SEQ_NUM, CRC_LS, CRC_MS K28.0, D0.0-D31.7, D0.0-D31.7, D0.0-D31.7	End of data frame control word. The SEQ_NUM field contains a number to identify duplicated or out of sequence data frames, broadcast frames or FCTs. The CRC_LS and CRC_MS fields contain a 16-bit CRC code computed by the sending node
SBF	Comma, SBF, BC, B_TYPE K28.7, D29.2, D0.0-D31.7, D0.0-D31.7	Start of broadcast frame control word. The VC field contains the number of the BC sending the data frame, and the field B_TYPE specifies its type
EBF	EBF, STATUS, SEQ_NUM, CRC K28.2, D0.0-D1.0, D0.0-D31.7, D0.0-D31.7	End of broadcast frame control word. The field STATUS informs about possible the delay of a broadcast frame. The SEQ_NUM field contains a number to identify duplicated or out of sequence data frames, broadcast frames or FCTs. The field CRC contains an 8-bit CRC code computed by the sending node
SIF	Comma, SIF, SEQ_NUM, CRC K28.7, D4.2, D0.0-D31.7, D0.0-D31.7	Start of idle frame. The SEQ_NUM field contains a number to identify duplicated or out of sequence data frames, broadcast frames or FCTs. The field CRC contains an 8-bit CRC code computed by the sending node
FCT	FCT, Multiplier/Channel, SEQ_NUM, CRC K28.3, D0.0-D31.7, D0.0-D31.7, D0.0-D31.7	It indicates that the channel indicated in the Channel sub-field has room for another complete data frame. The Multiplier is a 3-bit sub-field that identifies the value of the FCT. The SEQ_NUM field contains a number to identify duplicated or out of sequence data frames, broadcast frames or FCTs. The SEQ_NUM field contains a number to identify duplicated or out of sequence data frames, broadcast frames or FCTs. The field CRC contains an 8-bit CRC code computed by the sending node
ACK	Comma, ACK, SEQ_NUM, CRC K28.7, D2.5, D0.0-D31.7, D0.0-D31.7	It indicates that a data frame, broadcast frame or FCT was received without any error and in order. The SEQ_NUM field contains a number to identify duplicated or out of sequence data frames, broadcast frames or FCTs. The field CRC contains an 8-bit CRC code computed by the sending node
NACK	Comma, NACK, SEQ_NUM, CRC K28.7, D27.5, D0.0-D31.7, D0.0-D31.7	It indicates that a data frame, broadcast frame or FCT was received with an error or not in the correct order. The SEQ_NUM field contains a number to identify duplicated or out of sequence data frames, broadcast frames or FCTs. The field CRC contains an 8-bit CRC code computed by the sending node
FULL	Comma, FULL, SEQ_NUM, CRC K28.7, D15.3, D0.0-D31.7, D0.0-D31.7	It indicates that the Retry Buffer has become full. The SEQ_NUM field contains a number to identify duplicated or out of sequence data frames, broadcast frames or FCTs. The field CRC contains an 8-bit CRC code computed by the sending node
RETRY	Comma, RETRY, Reserved, Reserved K28.7, D7.4, D0.0, D0.0	It Indicates that a NACK was received, and that the sending node is transmitting data contained in the Retry Buffer

References

1. Dalal, S., Kumar, S., & Dixit, V. (2014). The OSI model: Overview on the seven layers of computer networks. *International Journal of Computer Science and Information Technology Research, 2*(3), 461–466.
2. Davalle, D., Nannipieri, P., & Fanucci, L. (2018). A novel parallel 8b/10b encoder: Architecture and comparison with classical solution. *IEICE Transactions on Fundamentals of Electronics, Communications and Computer Sciences, E101A*(7), 1120–1122.
3. Dinelli, G., Marino, A., Dello Sterpaio, L., Leoni, A., Fanucci, L., Nannipieri, P., & Davalle, D. (2020). A serial high-speed satellite communication codec: design and implementation of a SpaceFibre interface. *Acta Astronautica, 169*, 206–215.
4. European Cooperation for Space Standardisation. (2019). *SpaceFibre – Very high-speed serial link, ECSS-E-ST-50-11C*. European Cooperation for Space Standardisation.
5. Florit, A. F. (2017). Multi-lane communication. https://patentscope.wipo.int/search/en/detail.jsf?docId=WO2017134419
6. Koopman, P., & Chakravarty, T. (2004). Cyclic redundancy code (CRC) polynomial selection for embedded networks. In: *The International Conference on Dependable Systems and Networks, DSN-2004*, pp. 145–154.
7. Nannipieri, P., Marino, A., Fanucci, L., Dinelli, G., & Dello Sterpaio, L. (2020). The very high-speed SpaceFibre multi-lane codec: Implementation and experimental performance evaluation. *Acta Astronautica, 179*, 462–470.
8. Norte, D. (2011). Scrambling data signals for EMC compliance. In: *2011 IEEE International Symposium on Electromagnetic Compatibility*, pp. 471–475.
9. RapidIO.org. (2014). *RapidIO Interconnect Specification Version 3.1*. RapidIO.org.
10. Widmer, A. X., & Franaszek, P. A. (1983). A dc-balanced, partitioned-block, 8b/10b transmission code. *IBM Journal of Research and Development, 27*(5), 440–451.

Chapter 3
Building Blocks of a SpaceFibre Network: Examples Designs

3.1 The SpaceFibre CoDec

In this section, we present the architecture of the SpFi CoDec already shown in [4]. We aim to fully illustrate the architecture of the CoDec, alongside the design choice made and their impact on resource utilisation and power consumption. Consequently, the reader and potential system developer will be able to use the following text as a valid baseline to design its CoDec. The SpFi CoDec has three different interfaces allowing the host application to send and receive data over a SpFi communication link and to configure and monitor the status of each layer of the protocol. In particular:

- The *Configuration and Status* interface, which is generally connected to an upper layer with a register space for configuration and status control (the so-called Management interface). The latter can be implemented with a good degree of flexibility, depending on the application, therefore is not included in the following description.
- The *Physical* interface, transmitting and receiving SpFi encapsulated data words to/from the SerDes.
- The *Input/Output virtual channel* interface, transmitting and receiving SpaceFibre packets to/from the Host.
- The *Input/Output Broadcast Channel* interface, transmitting and receiving SpaceFibre broadcast messages to/from the Host.

The standard allows up to 32 couples of input-output VCs, to handle independent flows of information. The block diagram architecture of our SpFi CoDec is shown in Fig. 3.1. Besides the three interfaces aforementioned, we can observe the presence of three layers of the protocol stack: the Data-Link layer, the Multi-Lane layer (optional) and the lane layer. The rest is external to the CoDec.

The CoDec can operate with just one lane and do not include (bypass) the Multi-Lane layer or it can have multiple communication lanes, up to 16, operating

P. Nannipieri et al., *Next-Generation High-Speed Satellite Interconnect*,
https://doi.org/10.1007/978-3-030-77044-0_3

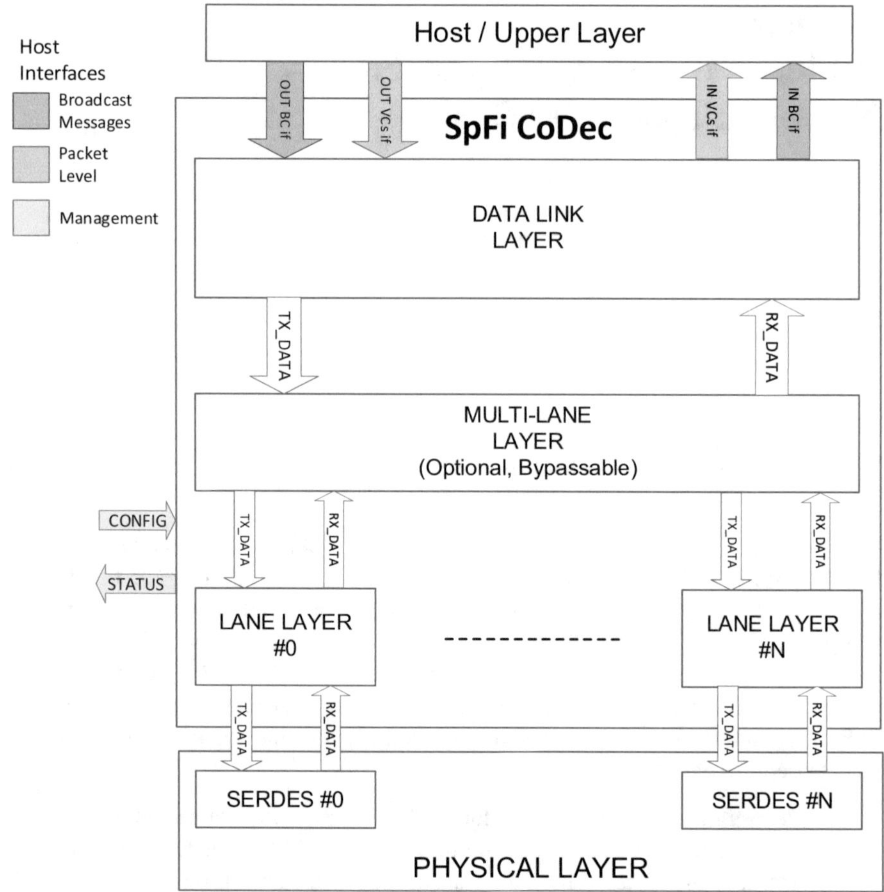

Fig. 3.1 SpaceFibre IP block diagram

in parallel; obviously, in case that there is more than one communication lane, the Multi-Lane layer needs to be instantiated and connected between the Data-Link layer and the lane layers to properly synchronise the communication. In the following paragraphs, a description of the architecture is given, focusing on the main features described in the SpFi standard.

3.1.1 Data-Link Layer

The architecture of the Data-Link layer is shown in Fig. 3.2. The OUTput Virtual Channel (OUT VC) interface is responsible for transmitting data packets from the upper layers (network/application). Each OUT VC shall be equipped with a buffer (with a minimum width of 256 words according to the standard).

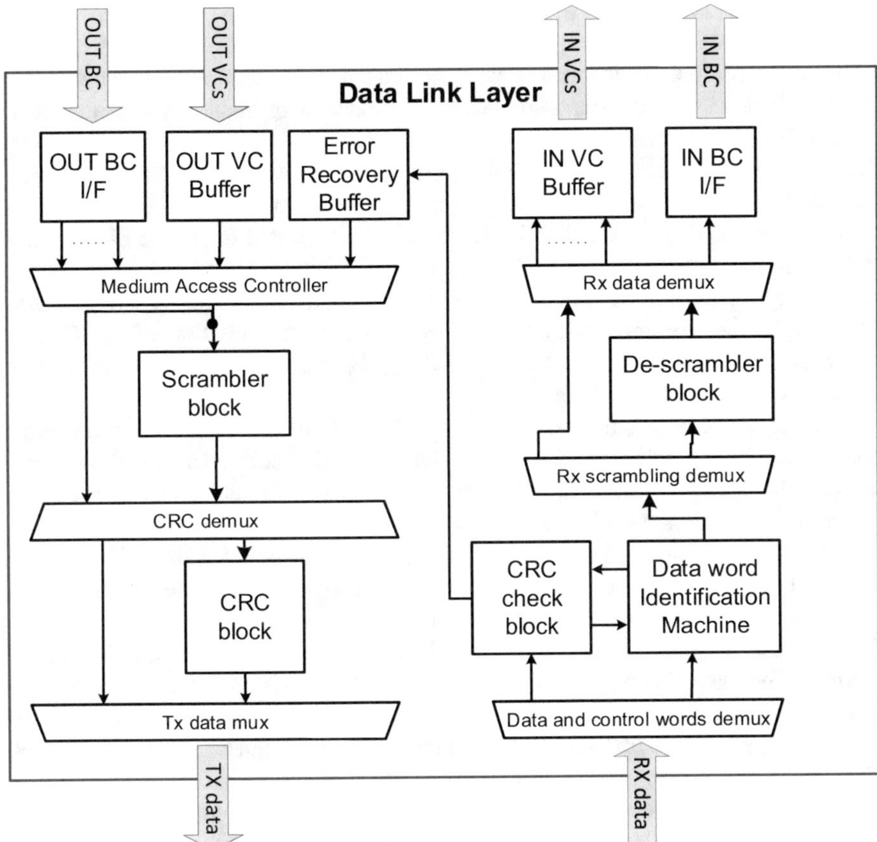

Fig. 3.2 SpaceFibre data-link layer block diagram [4]

The INput Virtual Channel (IN VC) interface is responsible for receiving data packets from the far-end of the link and for sending them to the upper layers. Each IN VC receives data from the far-end OUT VC with the same identification number, and it is equipped with a buffer, with a minimum size of 256 according to the standard. The OUTput Broadcast (OUT BC) interface is responsible for transmitting broadcast messages from the upper layer to the SpFi network. The INput Broadcast (IN BC) interface is responsible for receiving broadcast messages from the SpFi network, and for sending them to the upper layer. IN BC and OUT BC do not require a buffer, to preserve the low latency and high determinism feature of broadcast messages, which will be analysed in details in Sect. 5.3.2.

All buffers are implemented as asynchronous FIFOs (First-In-First-Out) to support clock domain crossing with higher layers. We have already mentioned that SpaceFibre is a fault-tolerant communication technology: it means that in case an error occurs, it can re-send data to avoid any data loss. The work published in [15] presents a study on all the possible algorithms, while the standard provides the exact algorithm to be followed for the scheduling. Furthermore, the MAC shall control the error recovery buffer, allowing to re-send corrupted data packet. This is a critical point of the entire CoDec and shall be designed carefully. The error recovery buffer is not fully detailed in the standard, thus there is a degree of freedom for the designer.

The MAC thus selects one of the OUT VC, the OUT BC or the Error Recovery Buffer for transmitting data across the link. It is also responsible for data framing (the first and the last word of a data frame or BC message are specific Data-Link control words). Optional data scrambling may be included to reduce electromagnetic emissions and finally, the CRC block calculates the cyclic redundancy check code. It is an error-detecting code [14], which is used to detect accidental changes to raw data that are common in the space environment due to radiation effects. CRC is inserted at the end of each data frame (CRC16), BC message (CRC8) and in specific control words (CRC8).

Thanks to the CRC, the far-end is capable of detecting communication error: indeed, the far-end calculates the CRC on its own on the received data and compares it with the received CRC. If they match, data is considered correctly received, otherwise an error is detected, and its request with a NACK data to be resent. To be fair, a NACK message and the consequent data retry request are generated also if the SEQ_NUM of the received packet is not consecutive to the previous one received, or if a lower level error (8B/10B parity error) occurred. The receiving side of the Data-Link layer operates specularly. It includes the IN VC interface and the IN BC buffers. The Data Word Identification state Machine (DWIM) searches constantly for Data-Link control words in the received data stream to identify if the far-end is transmitting a data frame, a BC message or serving a retry request. In parallel, the CRC block continuously computes the CRC code based on the incoming data stream. In case a mismatch with the received CRC is detected, a NACK request is generated and sent to the far-end of the link to the transmitting side. The descrambler block then de-scrambles received data packets if optional data scrambling is active. At this point, protocol control words are already been used and removed from the data stream and finally, data packets and broadcast messages are stored in the corresponding IN VC or IN BC.

Link Layer for the Multi-Lane Implementation

The inclusion of the Multi-Lane layer in the protocol stack determines major adjustments in the architecture of the Data-Link layer. Figure 3.3 shows the Data-Link layer architecture of the Multi-Lane version of the SpFi CoDec Intellectual Property (IP) core. The Data-Link layer designed performs the same operations described above for the single-lane implementation, but its architecture shall be modified to support Multi-Lane layer features. The *Retry buffer* and *VCs* shall be enlarged to provide a complete data row per clock cycle. Similarly, the *CRC block* and the *Scrambler block* shall perform their calculation over n words. The Switching Block IP core, or *SWIP*, is included only in the Multi-Lane version of the SpFi CoDec IP core. The *SWIP* does not interfere with Data-Link layer operations when all lanes are active, and it is responsible for handling data flow in case of one or more lane failures. Let us call the number of inactive lanes at a given time n_{fail}. In that

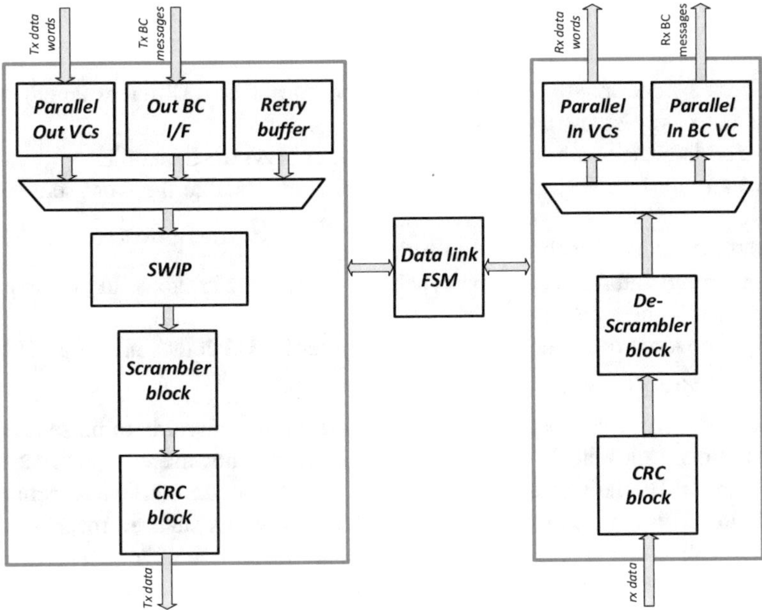

Fig. 3.3 Data-link layer architecture for the multi-lane version of the SpFi CoDec IP core [6]

case, the link could transmit only $n - n_{fail}$ words per clock cycle. However, the *Out VC* interface continues to provide n words at a time and the *SWIP* shall manage reading operation and synchronise Data-Link FSM to avoid the transmission of inconsistent data or even data loss. The *SWIP* shall handle a generic number of lane failures, guaranteeing a coherent data transmission, and its main features and architecture are described in the following section, according to our work presented in [5].

3.1.2 How to Handle Lane Failures: The SWIP Block

The SpaceFibre Multi-Lane layer shall detect real-time changes in the number of active lanes of the communication link, to preserve data integrity and speed up the link re-connection process. The SWIP takes as input a fixed-length data array and splits it on a variable number (equal or minor) of words equal to the number of lanes, which may change state in real-time, without any loss of data, preserving the initial order of the words. The block is designed to be inserted in the Data-Link layer of a Multi-Lane communication protocol, between the physical layers and the host interface, which is supposed to be implemented as a fixed-length asynchronous FIFO [2]. A series of general requirements have been derived from the standard for the SWIP block, as described in [5]. Let us consider N as the maximum number of parallel communication lanes of the system, M the actual number of active lanes and W_w the width in bits of the single word. The SWIP shall:

- process a fixed-length input data stream to map it on a different length output data path without any data loss.
- have an interface with the host able to transfer N-words in parallel.
- have an interface with lower layers able to keep separate the word sent to each lane.
- know the status of each communication lane.
- have a reset interface (may be synchronous or asynchronous, implementation-dependent) and a global enable.
- send the words over the link in the same order in which they are read out of the host FIFO.

The necessity of a block able to shape the number of words to be sent from a fixed number to a real-time variable number arises from these requirements. As input, the SWIP shall read NW_w-bit words to exploit the maximum achievable bandwidth. This implies that if $M < N$, the read words shall be mapped on the available active lanes, preserving data order and avoiding data loss. The literature lacks works compatible with such requirements. The circuit described in [23] is composed of two buffers and one multiplexer and appears to be not flexible in terms of the number of lanes to be handled. Another similar block is described in [3], but also in this case, not enough flexibility is provided, as the number of output words cannot dynamically changes and appears to be limited if compared with the SWIP. Our proposed solution instead can redirect a data path of N parallel words on M lanes, with M realtime variable.

An Example of SWIP Operation

The SWIP reshapes the data packet in case of one or more lane failures, allowing coherent data transmission. It selects M consecutive words, where M is the number of active lanes, and it assigns each one to the corresponding active lane. The proposed system was developed, integrated and tested with the Multi-Lane version of the SpFi standard [28]. To be clear, let us consider three lanes link composed of two active lanes and one inactive lane, shown in Fig. 3.4. The SWIP is composed of a two-register pipeline (Reg_0 and Reg_1) that contains N W_w-bit words. It shall compute the values of the signals *Fifo_read* and *index* to select two consecutive words per clock cycle (only two lanes out of three are active). *Fifo_read* is supposed to be the read enable of the host interface, and index is a counter that selects the words to be read. *Fifo_read* is set to *1* when there are no words to be read in Reg_1. In the first clock cycle, *(a)*, *index* is set to 0 and the system reads out W0/1. As W2 has not been read, no data is read out of the host FIFO and the shift between Reg_0 and Reg_1 is inhibited. After one clock cycle, in *(b)*, *index* value is 2, thus W2/3 are read across Reg_0 and Reg_1. No more data have to be read from Reg_1, thus in *(c)* a data row is read out from the host FIFO and Reg_0 is shifted in Reg_1. *Index* is set to 1 to read out W4/5. Finally, after one more clock cycle in *(d)*, the situation is back to the one of the beginning, with *index* set to 0, and W6/7 are read. As the last step,

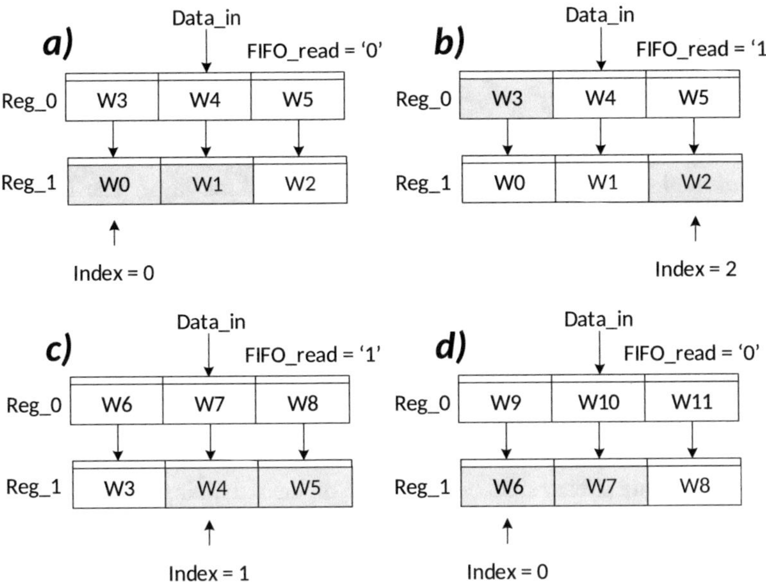

Fig. 3.4 Example SWIP operations [5]

the reshaping block shall assign the words selected by the FC FSM to the active lanes, using the signal *active_lanes* that indicates which lanes are active and which are not. In case one or more inactive lanes become active again, the SWIP will not be able to provide a valid word to each reconnected lane for one clock cycle. However, this is not critical for most Multi-Lane protocols (e.g. SpFi). Usually, a re-synchronisation mechanism is provided in case of lane re-connection with data stream being interrupted for one or more clock cycles. Thus, the signal *LL_read* will be set at "0", freezing the SWIP (*FIFO_read* set at "0" and *index* unchanged).

A Multi-Lane protocol may transmit both control words (e.g. the beginning and the end of a frame, error conditions, etc.) and data words. If we consider the SpFi standard, control words shall be replied on all the active lanes, while data words shall not. Moreover, if the number of words that composed a data frame is not a multiple of the number of active lanes, a special control word (named PAD) shall be inserted to form a complete row. Considering these requirements that are shared partially or completely with other Multi-Lane communication protocol, the FC FSM architecture has been designed.

3.1.3 Multi-Lane Layer

A schematic representation of the Multi-Lane layer is shown in Fig. 3.5. A summarised list of the requirements of the block follows. The *Tx_block* shall transmit data from the Data-Link layer to the lane layers. The *Rx_block* shall receive data from the lane layers and send them to the Data-Link layer. The *Lane Status* block shall inform in real-time the *ML-FSM* about the status of each lane. The *align_FIFO* shall buffer 3 or 4 words to ensure receiving row alignment. The *ML-FSM* shall:

- Allow the *Tx block* to receive words from the Data-Link layer and transmit them to the lane layers (one word per lane).
- Allow the *Rx block* to receive words from the lane layers and transmit them to the Data-Link layer.
- Inform the Data-Link about the status of each lane: in case of one or more lane failures, the data stream shall be redirected on the remaining active lanes.
- Check the alignment status of the link. Separate lanes may have up to a few clock cycle delays, respectively, due to different transmission mediums: this may lead to data row misalignment. The *ML-FSM*, thanks to a proper synchronisation protocol, compensates different delays reading out the *align_FIFO* words belonging to the same row.

The Lane layer maintains the same architecture in both single and Multi-Lane CoDec. The architecture of the SpFi Multi-Lane layer is shown in Fig. 3.5. The *Tx block* handles the transmission of data from the Data-Link layer to the lane layer, allowing data transmission only when all lanes are correctly initialised. The *Rx block* collects all data words received from lane layers, shapes them in data rows and passes them to the Data-Link layer only when lanes are correctly aligned.

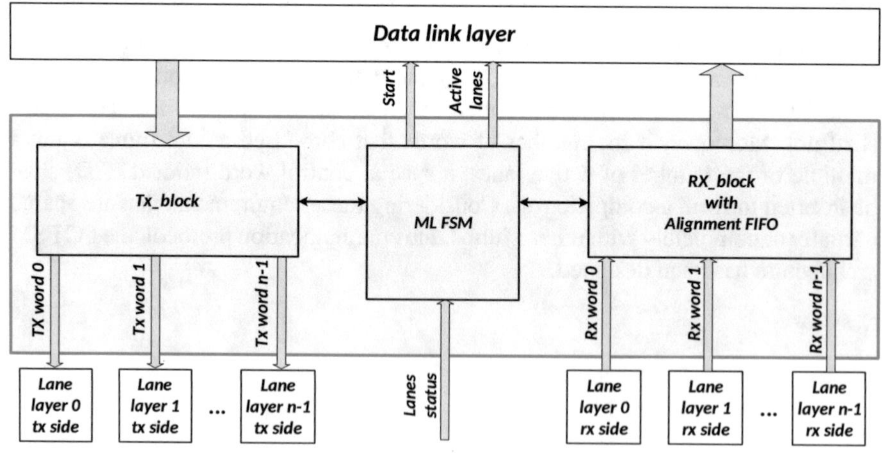

Fig. 3.5 Multi-lane layer block diagram [6]

The Alignment FIFO is responsible for the data alignment process. The *ML-FSM* manages the operations of all the blocks composing the Multi-Lane layer and promotes hot redundant lanes in case of a lane failure. Furthermore, it informs the Data-Link layer when the alignment process is complete and about the status of each lane (signal *Active lanes*), defining when the Data-Link layer can start to transmit and receive valid data words, through the signal *Start*. Even though the Multi-Lane support is probably the most complex feature of the SpaceFibre protocol, the Multi-Lane layer itself is not particularly complex. The real-time variation of active lanes is handled at Data-Link layer level, therefore the biggest share of the introduced complexity is within that block. Please note that this is not a direct requirement of the standard, which just describes at a high-level the layer requirements, which are then elaborated and mapped in real architecture.

3.1.4 Lane Layer

The Lane layer is responsible for establishing communication across a SpFi link. The architecture of the Lane layer is shown in Fig. 3.6. All the operation carried

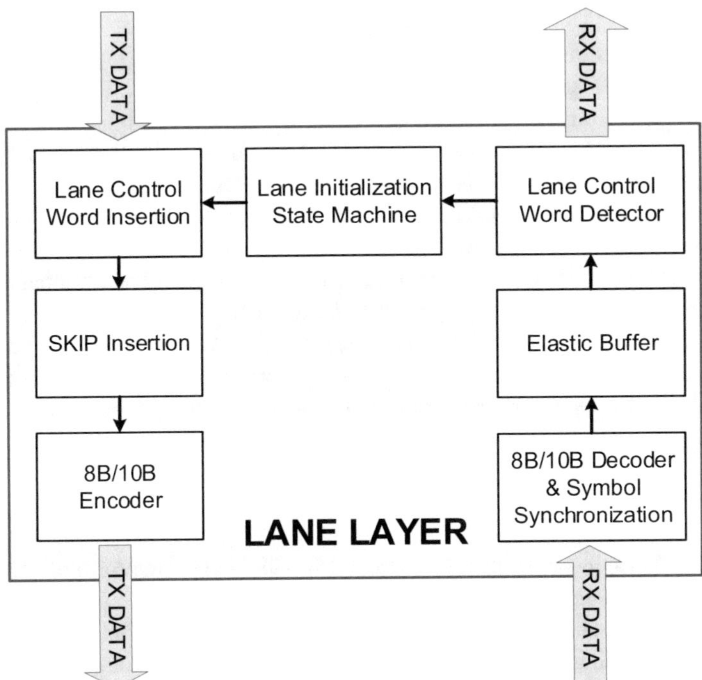

Fig. 3.6 SpaceFibre lane layer block diagram

out by the Lane layer pass through the control mechanism provided by the lane initialisation FSM, which has the role to align and synchronise input and output data streams.

The synchronisation between the two ends of the link is obtained through Lane layer control words exchange. The Lane layer is also responsible for compensating differences that may arise between the data signalling rate of the two ends of the communication link. This may happen due to slight differences in the local clock of the two ends of the link. The standard supports a difference up to 100 ppm. Therefore a receive Elastic Buffer shall be included to accommodate small differences between the two clocks, but also to interface two separate and asynchronous clock domains. The elastic buffer recognises a special SKIP word in the data stream and, if the buffer is more than half-full, the SKIP word is effectively skipped. If, on the other hand, the buffer is less than half-full, the SKIP word is not skipped. One SKIP word is transmitted every 5000 words. Finally, the Lane layer is responsible for encoding data and control words into symbols through 8B/10B Encoder during data transmission, and for decoding received symbols into data or control words through the 8B/10B decoder. In the transmission data stream, Lane layer control words are inserted at first together with the SKIP words under the supervision of the lane initialisation FSM; then, the data streaming encoded and passed to the SerDes. On the receive side, raw data is received from the SerDes, decoded, passed through the elastic buffer and finally, after that Lane layer control words are removed, send to the upper layer.

For what concerns the SerDes interface (e.g. the Physical layer interface), we shall consider that many SerDeses are available on the market, also depending on the silicon technology: there are SerDes circuits which just serialise and deserialise the data stream, while there are ones which implement the full stack of the Lane layer excluded the lane initialisation FSM. A key value for an HDL IP is to be as much configurable and flexible as possible. Therefore the SerDes interface of the Lane layer should be configurable at various levels of the Lane layer data stream: 8B/10B, elastic buffering and the width of the data patch shall be optional and user-configurable, to maintain compatibility with all existing devices.

The adopted 8B/10B encoder/decoder is the one presented in [25] and described in the following section, which employs an innovative parallel approach so that is possible to save logical resources and reduce power consumption.

3.1.4.1 8B/10B Parallel Encoding/Decoding

8B/10B encoding has been firstly developed in 1983 [31]. This form of encoding is largely used in several communication protocols, such as PCI Express, Serial ATA, USB3.0, Fibre Channel and many more, including high-performance communication protocols such as SpaceFibre, the future standard for communication on-board of spacecraft [1]. 8B/10B encoding takes an 8-bit symbol as input together with its control bit and produces a 10-bit character for transmission. The popularity of this technique is due to its advantages against direct transmission:

- The transmitted data stream has roughly the same number of zeros and ones, resulting in a zero DC bias, thus enabling AC coupling.
- The number of transitions is enough (maximum 5 consecutive zeros and ones) to enable the recovery of the bit clock with a Phase-Locked Loop (PLL).
- All characters are transmitted with 10 bits, thus the transmission rate is constant.
- 10 bit codes have 1024 possible values: there are values left for control characters and values that shall not be used, giving the possibility to easily identify possible link errors.
- The current running disparity, which will be defined below, can be only +1 or −1, another value indicates an error, also resulting in easier link error detection.

8B/10B encoding uses only characters that contain either 5 ones and 5 zeroes, 6 ones and 4 zeroes or 6 zeroes and 4 ones, to generate zero DC bias: when a character with 5 zeroes and 5 ones is sent the produced DC bias is 0, but this would slowly increase with a series of character with more ones than zeroes or vice versa. To avoid this, each character, except the ones with an equal number of zeroes and ones, has a double form of encoding, one with 6 zeroes and 4 ones and the other with 6 ones and 4 zeroes. The transmitter keeps track of what has been sent in the previous character with the Current Running Disparity (CRD), so that after a character with more zeroes a character with more ones will be sent, and vice versa, resulting in the desired zero DC bias. For the encode table, please refer to [31]. The single symbol encoder architecture, specified in [12], is reported in Fig. 3.7, while the classical architecture two symbols encoder, specified in [29], is reported in Fig. 3.8.

8B/10B encoders and decoders are attached to SerDes circuits, which perform high-speed serialisation and de-serialisation of data. Typically a SerDes takes more than one parallel 10-bit characters to be serialised and then it sends them over the serial link. It is very common in particular to have 20-bits SerDes, consequently,

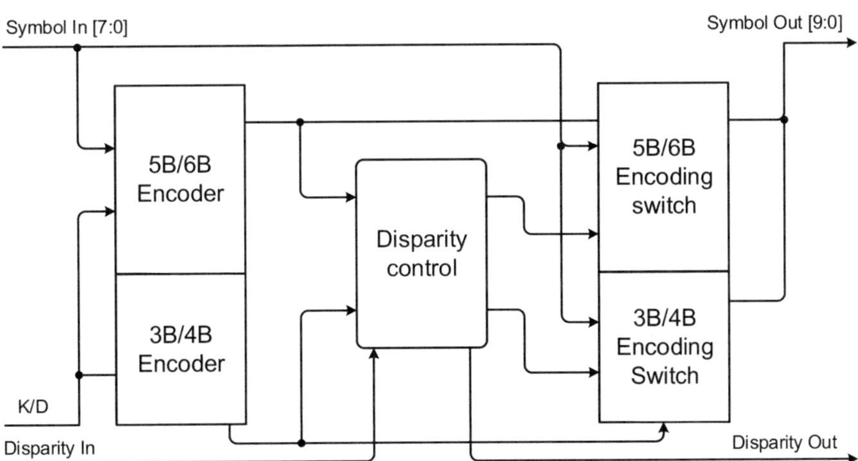

Fig. 3.7 Generic 8B/10B encoder scheme [12]

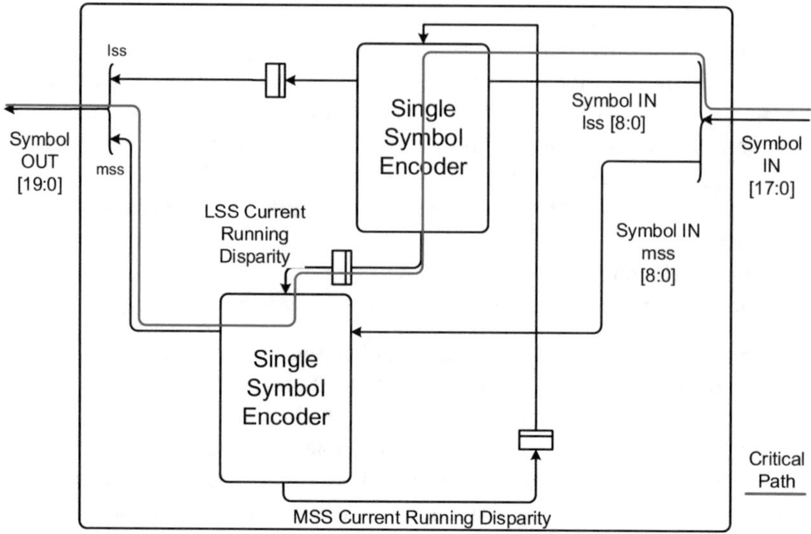

Fig. 3.8 Two symbols pipelined 8B/10B encoder

the symbols must somehow be processed simultaneously. This is done inefficiently with the classical architecture shown in Fig. 3.7 since the Lowest Significant Symbol (LSS) encoder must produce its CRD, which will be used by the Most Significant Symbol (MSS) encoder as input. To produce coherent encoded symbols, either a very long logic path is needed or the encoder must be clocked N times faster than the rest of the circuit, where N is the number of parallel symbols produced by the SerDes. In case of high-speed requirements, it is necessary to insert a pipeline register, as done in [1] to shorten the critical path, at the price of increased latency. In Fig. 3.8 the architecture of a generic two symbols encoder is shown, with the optional pipeline registers placed. The red dotted line shows the critical logic path in case that no pipeline register is inserted.

The usage of pipeline registers results in a higher number of logic resources used, and the latency increase. A possible way to handle this problem is to introduce a parallel encoder, able to elaborate N symbols per clock cycle. The architecture of such a system is hereby presented, and its performance is compared with one of the classical pipelined systems. The parallel architecture proposed is shown in Fig. 3.9.

The idea is to predict the parity signal required by the single encoder. This is done by analysing input symbols. Part of the logic, in the form of Look Up Tables (LUT), of the single symbols encoder must be replicated within the parity ahead calculator block. However, carefully analysing the encoding table, it is possible to significantly reduce the complexity of this block. The idea behind the parity ahead calculation is explained in Fig. 3.10.

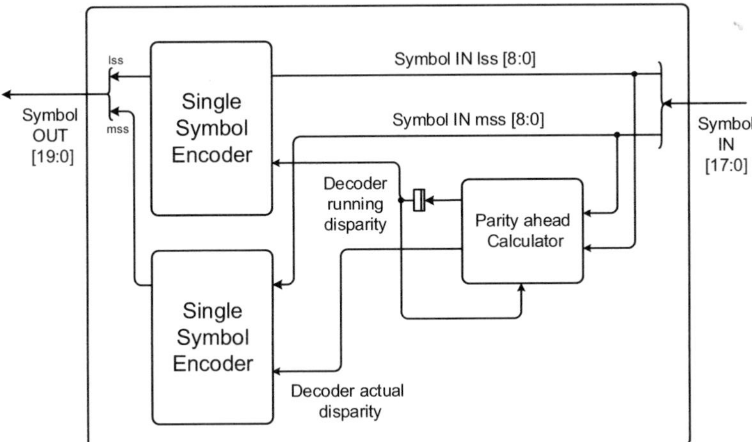

Fig. 3.9 Two symbols parallel 8B/10B encoder

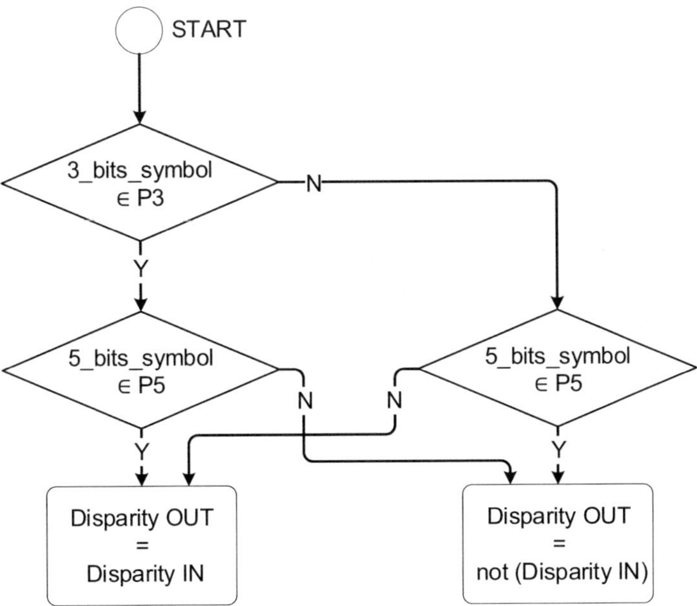

Fig. 3.10 Parity calculation diagram

Let us call *Disparity IN* the known disparity of the $(K-1)^{th}$ symbol and *Disparity OUT* the disparity of the Kth symbol, which we want to calculate. At the input of the parity calculator, we will have both the uncoded 8-bit symbol and *Disparity IN*. The 8 bits symbol is composed of a 3-bit symbol, 3_bits_symbol, and a 5-bit symbol, 5_bits_symbol. From the encoding table, we observe that the sub-symbols must belong to one of the following sets:

- 3_bits_symbol
  - Set P3: The set of the 3 bit symbols whose encoded version has neutral parity.
  - Set D3: The set of the 3 bit symbols whose encoded version has ±2 parity.
- 5_bits_symbol
  - Set P5: The set of the 5 bit symbols whose encoded version has neutral parity.
  - Set D5: The set of the 5 bit symbols whose encoded version has ±2 parity.

Hence the parity ahead calculation block has just to keep track of which sets every possible input belongs, then apply two conditional statements. This can be done in cascade for N parallel encoders; the higher the number of parallel symbols is, the longer the critical path to compute all the parity in this combinational block will be. A comparative analysis has been performed between the parallel architecture proposed in Fig. 3.9 and the classical one of Fig. 3.7, with a pipeline stage. The comparative analysis has been performed synthesising both the architectures on a 65 nm CMOS standard cell technology. Please note that the routing factor has not been taken into account, as only a preliminary synthesis has been performed. The results have been compared in terms of maximum achievable clock frequency, the total area occupied in terms of μm^2 and Kgate, total dynamic power and throughput. In particular, two separate syntheses have been performed, one aiming at achieving the highest clock frequency possible and one aiming at reducing the area occupied, with a given target clock frequency f_{clk} = 312.5 MHz, which is the operating frequency of a high-performance communication protocol such as SpaceFibre. The throughput of the system is defined as the number of symbols encoded per clock cycle, where f_{clk} is the frequency at which the SerDes consumes N parallel symbols, with $N = 2$.

From the results shown in Tables 3.1 and 3.2 we can observe that:

Table 3.1 Pipelined [12] and parallel encoders: timing effort

	Pipeline configuration [12]	Parallel configuration
Area [μm^2]	7843	7405
Area [KGate]	3771	3560
Max freq. [GHz]	1.1	1.05
Dynamic power [mW]	3.96	2.89
Throughput [symbols/clk cycle]	1	2

Table 3.2 Pipelined [12] and parallel encoders: target frequency 312.5 MHz

	Pipeline configuration [12]	Parallel configuration
Area [μm^2]	7121	5924
Area [KGate]	3423	2848
Dynamic power [mW]	1.1	0.58
Throughput [symbols/clk cycle]	1	2

- The logic path created with the combination of the parity ahead calculator (Combinational logic) and the single symbol encoder is not affecting the maximum clock frequency, which is just 4.5% lower. This is because the parallel configuration enables a not-negligible simplification in the Single symbol encoder.
- The parallel configuration area is lower in both the cases analysed, with a reduction of 17% in the fixed frequency synthesis. This is due to the lower number of registers used in the second configuration, together with the simplification mentioned above. The area reduced this way completely exceeds the area added for the parity ahead calculator block.
- Consequently the dynamic power consumption of the parallel configuration is lower in both cases.
- The throughput is doubled with the parallel configuration: we can encode two symbols per clock cycle by predicting with a combinational block the input parity, without the need to wait for the previous symbol to be encoded and its parity calculated. On the contrary, the configuration of Fig. 3.8 needs two clock cycle to produce two encoded symbols, thus it has to be clocked with $f'_{clk} = 2f_{clk}$, which reduces the throughput of a factor 2. Please note that it is not possible to insert a pipeline register on the second encoder input, to double the throughput, (increasing the latency) as this will break the synchronisation loop of the CRD.

The two configurations are comparable for what concerns the maximum clock, frequency, area and dynamic power consumption; the parallel encoder has better performance in all these fields except for the maximum frequency. The great advantage of the parallel configuration is that the throughput is doubled without deterioration in frequency and area figures, which is a tremendous improvement. The procedure applied to this circuit with $N = 2$ applies to a generic N, in particular with $N = 4$, which may be another interesting case. However, it is expected that the complexity of the parity ahead calculator block will increase as N increase.

3.2 The SpaceFibre Bus Functional Model and Verification Environment

A Bus Functional Model (BFM) is a hardware model, written in non-synthesisable high-level languages, of a circuit with external bus interfaces. In our case, we can imagine the SpaceFibre host interface to be our bus. For any designer willing to develop a complex hardware IP such as a SpaceFibre interface, it is almost mandatory to have such a system, acting as a golden reference model, to verify the complete functionality of the developed circuit independently from the HDL. Moreover, a BFM model can also be used as a simulated conformance tester, at the moment of integrating external SpaceFibre IPs in a more complex system. In the following, we will detail a possible architecture and working principles of a SystemVerilog based SpaceFibre BFM hardware verification environment, as presented in [4]. The approach of employing a golden reference, to be developed as an independent high-level model of the CoDec, demonstrated to be effective in the identification of bugs, misunderstandings and ambiguities in the standard requirements processing. Whether you decide to use the BFM SpaceFibre model to develop your system or to test third-party IP conformance to the standard, you need to perform an appropriate hardware verification through the execution of a tests suite. If the verification campaign objective is to test standard conformance, then the execution of the test plan and the report produced is already sufficient as the output product of the verification. on the other hand, if the objective is to verify a newly developed SpaceFibre hardware CoDec, then it is advisable to reach a 100% simulation coverage score in (1) statements coverage; (2) branches coverage; (3) FSM coverage and (4) conditions coverage. This verification methodology has been adopted even if it does not present the state of the art, which is currently represented by formal verification [32]. Even if it is a sub-optimal solution, it is a common approach in space-related HDL design verification: being the system significantly complex, it would require considerable manpower to be formally verified, which is not usually available due to the limited market for these products. This approach is considered the baseline in the pace community; indeed, the adopted methodology is compliant with the design flow described in the ECSS standard for ASIC and FPGA development [9]. Our verification environment was built using the Universal Verification Methodology (UVM) [16] framework for SystemVerilog. It represents the state of the art in the digital design verification field, consisting of a set of built-in classes, macros and libraries allowing rapid and robust development of the verification environment. One of the greatest advantages of UVM is the possibility to exploit the Transaction Level Modelling scheme: all the TestBench (TB) logic are developed at a transaction level instead of at the pin level. Note that in UVM, a transaction is represented as a SystemVerilog class. This greatly simplifies the verification environment development and reduces test-bench and test case design time. It is out of the scope of this text to provide more details about the UVM framework, which the reader can find in [16]. It is recommended to follow a similar approach in the design of a BFM and of a verification environment for a SpaceFibre interface: the higher level the environment is encapsulated, the easier will be to write and create complex test cases. The fact that the verification environment was built as a BFM of a SpaceFibre CoDec means that a SpaceFibre module was developed in SystemVerilog according to the SpaceFibre standard [10] and that model was connected at PIN level to the RTL IP Core Device Under Test (DUT). In this way, the functionality of the DUT is tested cycle-accurately, with visibility of each RTL node, while specifying the test cases at a high-level.

3.2.1 Verification Environment Architecture

Figure 3.11 shows a generic BFM scheme. The main components of the verification environment are detailed in the following.

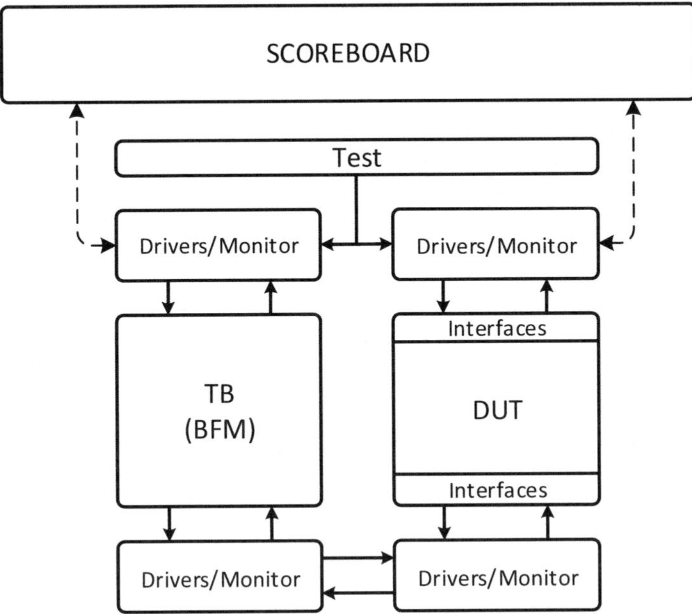

Fig. 3.11 SpaceFibre verification environment scheme with bus functional model

Bus Functional Model

The SystemVerilog BFM is the core of the verification environment. It is a SystemVerilog implementation of a SpaceFibre interface and it is used to simulate the communication of the DUT with an external CoDec. The model shares all the configuration parameters with the DUT (for example, the number of virtual channels). The model is realised to withstand all the SpaceFibre standard requirements; however, the fact that it does not need to be written in synthesisable HDL makes the design process of the model much faster. It is an important good practice that the model of the SpaceFibre CoDec is designed independently, possibly by different designers, from the Hardware Description Language (HDL) SpaceFibre CoDec. This will strengthen the verification process.

Interfaces

The DUT is connected to the rest of the environment through SystemVerilog interfaces. They are used to group synchronous signals, depending on their functions. The interfaces are the same already specified for the SpaceFibre CoDec:

- Data-Link layer: there is one network layer interface for each virtual channel.
- Physical layer: there is one Physical layer interface representing the serial bus for the connection with the SystemVerilog model.
- Management layer: there is one management layer interface for the status and control information. This interface is used also to send the reset signal.

Drivers and Monitors

Drivers and monitors are the components realising the translation between the pin level world of the DUT and the transaction level world of the TB. The role of drivers is to take a transaction coming from the SystemVerilog model, to drive the corresponding interface signals to send that transaction or to read a sequence of signals from an interface and build a transaction to send to the model. Monitors are passive elements that perform correctness checks on the interface signals. For each interface, there are two drivers, one for the communication model-DUT and one for the communication DUT-model, plus one monitor.

Scoreboard

The scoreboard keeps track of all the statistics of the tests, including the number of errors that occurred, the number of packets sent and arrived and their latency. The controlling idea of the scoreboard is to monitor all the received data words at the two ends of the communication link and compare them with a list of expected received data words, computed independently.

Test

The above-mentioned components represent the architecture of the verification environment, which is to be used to run tests. The test case is the highest level component of the verification environment, which on one side instantiates the entire verification environment, and on the other provides high-level stimuli to it. The test environment is developed in such a way that it is possible to run single test cases or aggregate test suites. For each test case, a different test file is created. The verification environment produces two outputs for each test file: a coverage file, containing the coverage information generated by the simulation tool and a result file, containing PASS/FAIL information and the detailed description of errors that occurred. The verification environment produces a coverage file for each executed test case and can merge separate test coverage files into a unique database. The advantage of this approach is that the same infrastructure can be used at different steps of the verification process; in the early verification stages, basic tests will be launched to verify the lowest level mechanism of the system (e.g. communication between DUT and BFM works correctly). Going on with the verification, more complex tests will populate the test plan. In the end, it is possible to obtain a heterogeneous set of tests, developed to fully verify the functionalities of the CoDec, which can be run separately or in parallel to reduce the execution time. Once all the test have been executed, acquired coverage files can be post-processed and merged, to evaluate the quality of the overall verification effort.

3.2.2 Verification Plan

The verification plan is the core element of the overall testing system: it describes one by one all the test cases used to verify the IP core. These tests have been written varying both configuration parameters and creating different situations during simulation time to evaluate all the corner cases. The objective is to reach a 100% test coverage, in the lowest simulation time possible. The verification plan shall be designed carefully keeping in mind the protocol requirements to be verified and the high-level instruments which the verification environment provides to the test engineer. In each test, several parameters are defined for DUT and BFM CoDecs:

- N VCs: the number of virtual channels used for the test. This value must be equal between the TB and the DUT. Allowed values are [1, 32].
- N Lanes: the number of communication lanes to be instantiated and connected within the BFM and DUT. This value must be equal between the TB and the DUT. Allowed values are [1, 16].
- N Hot Redundant Lanes: the number of the hot redundant lanes to be instantiated and connected within the BFM and DUT. This value must be equal between the TB and the DUT. Allowed values are [0, 15] and shall be lower than N Lanes.
- N Tx/Rx Only Lanes: the number of the Tx/Rx only lanes to be instantiated and connected within the BFM and DUT. This value must be equal between the TB and the DUT. Allowed values are [0, 15] and shall be lower than N Lanes, as a SpaceFibre link requires at least one active bi-directional lane.
- Timeslot: the timeslots used by each virtual channel. The number of timeslots is always set equal to the number of virtual channels. A VC can transmit in a certain timeslot (1) or not (0).
- Priority: the priority level of each virtual channel. Allowed values are [1, 15].
- Expected bandwidth: the expected bandwidth for each virtual channel. Allowed values are [0, 100].
- Expected BC bandwidth: the expected bandwidth for the broadcast messages. Allowed values are [1,100].

All the parameters listed here above are set according to the specification of the SpaceFibre standard. Let us define also the concept of actual bandwidth:

- Actual bandwidth: this value represents the rate at which data are written into the virtual channel by the application. Depending on the situation and the QoS used, the virtual channel transmission buffer can become full. Allowed values are [1, 100].
- Actual BC bandwidth: this value represents the rate at which data are written into the broadcast buffer by the application. Depending on the situation, the broadcast transmission buffer can become full. Allowed values are [1,100].

As mentioned before, the verification plan is built incrementally, to intensively test from the basic functions of SpaceFibre (simple communication) to the most complex ones (corner cases, mixed QoS). The main functions under test are:

- Basic communication: this set of tests investigates a simple communication between the model and the DUT, also using broadcast messages.
- Error Injection: this set of tests stimulates the FDIR mechanism of SpaceFibre, injecting errors on the link.

- Broadcast Burst: this set of tests aims to stress the DUT with a burst of broadcast, also injecting errors.
- Quality of Service—Basic: this set of tests assesses each QoS mechanism of SpaceFibre (priority, bandwidth reservation, and time scheduling) separately.
- Quality of Service—Mixed: this set of tests assesses the three QoS mechanisms mixed.
- Multi-Lane—basic: this set of tests assesses the basic functionalities of the Multi-Lane layer
- Multi-Lane—advanced: this set of tests assesses advanced functionalities of the Multi-Lane layer, combining the use of more than one lane, Tx/rX only lane s and also hot redundant lane, with error injection and lane disconnection.
- Reset: this set of tests stimulates the reset signals during the normal CoDec operations.
- Clock domains: this set of tests involves the asynchronous interfaces of the DUT.

All the tests in the test plan were successfully executed on our designed SpaceFibre CoDec IP core. A coverage of 100% was reached for statements, branches, finite state machines states and finite state machines transitions, fully verifying the Register Transfer Level (RTL) description of the SpaceFibre CoDec according to [9].

3.3 The SpaceFibre Router

A router is a networking device that forwards data packets between separate endpoints belonging to the same network. SpaceFibre may be used to connect several instruments to the on-board mass memory through routing switches. This section will introduce the design and implementation process of a SpaceFibre routing switch with the multicast feature.

3.3.1 SpaceFibre Routing Switch Architecture

3.3.1.1 Architecture Overview and Main Features

An overview of the Register Transfer Level (RTL) architecture of the proposed Routing Switch is shown in Fig. 3.12, as presented in [27]. The Routing Switch itself does not directly comprise the SpaceFibre ports, but it has a FIFO interface within the port manager to connect with them. The Port Manager simplifies the interaction between the external SpaceFibre port and the internal logic. In particular, it keeps track of the input and output state of each VC in a port through the two FSMs represented on the left side of Fig. 3.13. The "Input Port to Switching Logic" FSM (on the left) manages the incoming packets. When a new packet is received by a

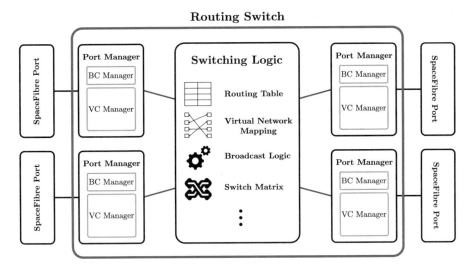

Fig. 3.12 Overview of the Routing Switch architecture [27]

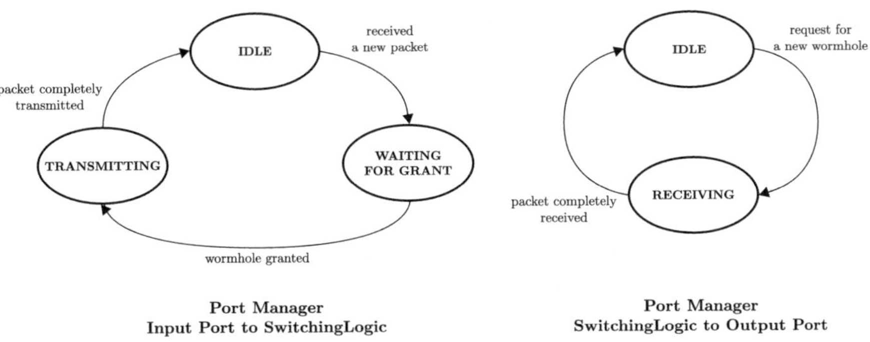

Fig. 3.13 Port manager FSMs for VCs handling

<Port, VC>, a request for a new wormhole connection is issued to the Switching Logic and the FSM goes from IDLE to WAITING FOR GRANT state. The FSM also specifies the address requested by the incoming packet.

This request may take some time before been accepted, depending on the state of the output ports which can be already busy with some previous wormholes. However, the SpaceFibre FCT system guarantees that no data in the IN VC interface will be lost. Once the request is accepted, the FSM moves to the TRANSMITTING state, opens the wormhole and stays in this state until the packet is fully transmitted, up to and including the EOP (or EEP) marker. The "Switching Logic to Output Port" FSM is a bit simpler. The FSM remains in the IDLE state until the Switching Logic notifies it that a new request has been issued. At this point, the FSM moves to the RECEIVING state until the whole packet has been received. During this time, the <SpFiPort, VC> is busy and cannot accept new wormholes. Apart from the Port

Manager, all the functionality of the SpaceFibre Routing Switch is implemented inside the Switching Logic. In particular, it is responsible for:

- decoding the path or logical address of a new request;
- decoding the VN of both the input and output port;
- implementing a scheduling policy in case of concurrent input requests for a busy port;
- implementing the logic for all the mechanisms foreseen by the SpaceFibre standard, such as Multicast, Group Adaptive Routing (GAR) and VC Timeout;
- implementing the Broadcast forwarding mechanism.

In the following sections, these features are described.

3.3.1.2 Routing Table

When a new request is issued by an input <Port, VC>, the corresponding Port Manager indicates also the requested address, which can be either a path or a logical address. In case it is a path address (address < 32), the output port coincides with it and there is nothing else to do. In case it is a logical address however, the Switching Logic has to decode the associated output port by accessing the Routing Table (Fig. 3.14).

The Routing Table, for each possible logical address (32–254), contains:

- a 32-wide bit vector of the output ports. Because a certain logical address can be associated with more than one port (in the case of Multicast or GAR), a bit vector is the most efficient implementation. If the n-th bit is a 1, then the n-th port is associated with that logical address (it is part of the output ports set);
- a flag (1 bit) specifying if the output ports set is a Multicast or a GAR set. When only one output port belongs to the output set, this flag is ignored;
- a 32-wide bit vector for the header deletion. According to the SpaceFibre (and SpaceWire) standard, when a packet using a logical address is forwarded, the Routing Switch can remove the first byte of its header to support Regional

	Output ports	Multicast/ GAR	Header deletion
LogicalAddr 32 →	0100...0000	/	0000...0000
LogicalAddr 33 →	0001...0000	/	0000...0000
LogicalAddr 34 →	0111...0000	GAR	0110...0000
	⋮	⋮	⋮
LogicalAddr 255 →	0000...0101	MUL	0000...0000

256-32=224

Fig. 3.14 Routing Table as implemented in the Switching Matrix [27]

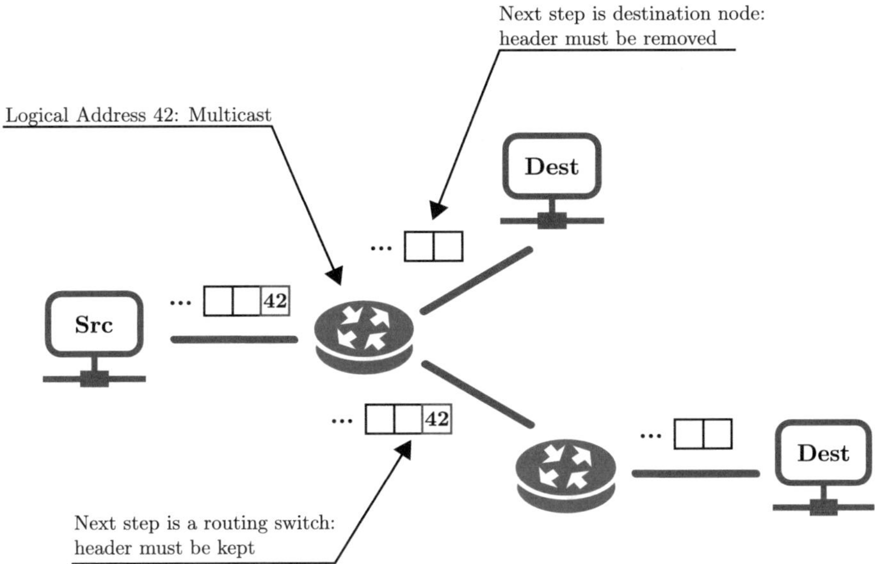

Fig. 3.15 Corner case in which the header deletion is applied on a per-port basis for the same logical address

Addressing or destination nodes that do not expect the first byte to be the address. Note that the use of Multicast or Group Adaptive Routing can create more complex scenarios in which the header must be removed or kept depending on the output port to which it is forwarded, even if it is associated with the same logical address. Figure 3.15 shows an example of that, where logical address 42 corresponds to a Multicast set in the first Routing Switch along the path. Because of the peculiar network topology, and assuming that the destination nodes expect to receive a packet without the address, the header must be removed while forwarded to the top node and kept while forwarded to the bottom next Routing Switch.

After the decoding of the output port (or ports), the VN decoding step must be done.

3.3.1.3 Virtual Network Mapping

As already said, only VCs belonging to the same VN can be connected. However, VN mapping is not static but can be dynamically changed. This means that, from an RTL standpoint, each VC of each other port must be connected to each VC of each port, even if most of these connections are not used during the operation, greatly increasing the complexity of the design. When a request for a new wormhole arrives

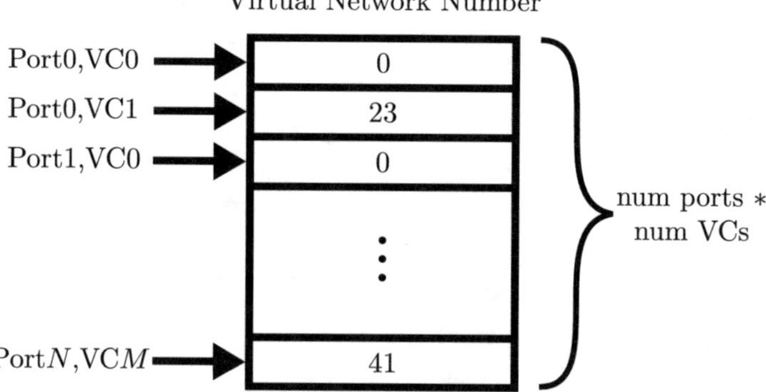

Fig. 3.16 Virtual Network Table containing the VN associated to each <Port,VC>

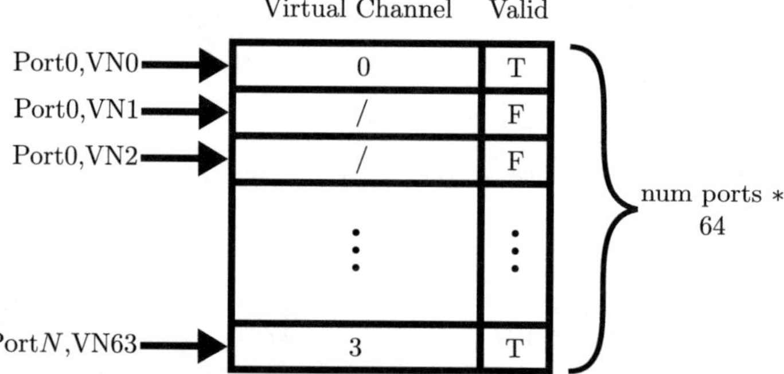

Fig. 3.17 Inverse virtual network table, used to retrieve the VC associated to a VN

and the output port (or ports) has been found, the VN of the input VC must be decoded. For that, a Virtual Network Table (Fig. 3.16) is implemented.

The Virtual Network Table allows mapping each <Port, VC> to a specific VN number. The next step is to find, for each output port of the output ports set, the VC associated with the same VN.

This is done through the Inverse Virtual Network Table, shown in Fig. 3.17. It maps each <Port, VN> to the corresponding VC, avoiding multiple accesses to the previous table. Because it is possible that some VNs are not associated with any VC (while a VC is always associated with a VN), this table contains also a valid flag to indicate if the corresponding entry is valid or not. These steps are summarised in Fig. 3.18, which underlines the accesses to the two tables. This decoding process is iterated for each output port, at the end of which the input <Port, VC> is put in a waiting state until all the output <Port, VC> requested are free to accept a new wormhole.

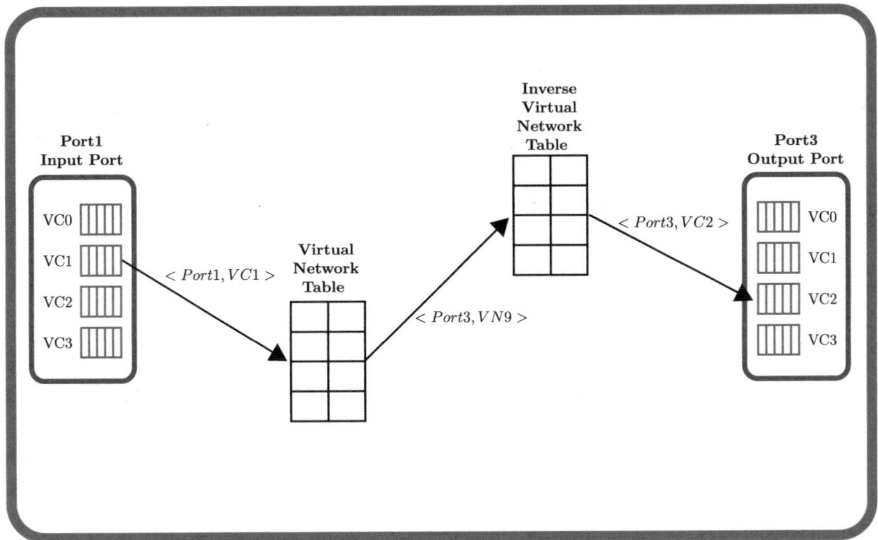

Fig. 3.18 Steps to find the VC in the output port belonging to the correct VN [27]

3.3.1.4 Scheduling Policy

When multiple input ports are waiting to obtain an output <Port, VC>, a scheduling policy is necessary to decide who goes first. The SpaceFibre standard does not specify any scheduling policy, leaving the choice to the implementer. The scheduling policy we proposed and developed is a Round Robin (RR) for each output <Port, VC>, granting the maximum level of fairness among the ports and avoiding deadlock situations. The entity implementing it is the Round Robin Arbiter (RRArbiter), and there is one of it for each output <Port, VC> to have that policy implemented reasonably. The role of that arbiter is to keep track of the requests from the input ports and giving, as output, the unique input port that is allowed to get the next wormhole.

3.3.1.5 Switching Matrix

The so-called Switching Matrix is the core element of the Routing Switch and it is the entity responsible for the actual forwarding of data from the <inputPort, inputVC> to one or multiple <outputPort, outputVC>. In this Section, only the VC data is taken into account, while the Broadcast forwarding is described in Sect. 3.3.1.6. To summarise, the Switching Matrix comprises a combinational block and a sequential block.

Combinational Logic

The combinational logic connects the input and output signals of each <Port, VC> so that the data can be transferred. There are no pipeline stages in that connection (however, everything is already registered inside the correspondent Port Manager). Consider that a <Port, VC> is always bi-directional and the names “input” and “output” only refer to the transfer direction taken into account in a specific moment.

3.3.1.6 Broadcast Logic

The Broadcast service is one of the most innovative features of SpaceFibre, greatly extending the limited TimeCode (and now Interrupt) mechanism of SpaceWire. The format of a Broadcast message is shown in Fig. 3.19, and it carries a payload of 8 bytes delimited by two SpaceFibre control words, are the Start Broadcast Frame (SBF) and the End Broadcast Frame (EBF), respectively. These control words contain a lot of information, the most important are:

- Channel, which uniquely identifies the source node generating the message. This practically limits the number of nodes that can generate Broadcast messages in a network to 255;
- Type, identifying the type of the message. This field allows to use the Broadcast service to implement multiple protocols (up to 255), differently from SpaceWire where there are only TimeCodes and Interrupts;
- LATE, a flag that is set by a Routing Switch or by an end-node in case of the message experiences some kind of delay before being sent, for example, it is subject to a retry operation caused by errors on the link;

Type	**Channel**	**SBF**	**Comma**	**SBF Control Word**
D3	*D2*	*D1*	*D0*	
D7	*D6*	*D5*	*D4*	
CRC	**SeqNum**	**Late**	**EBF**	**EBF Control Word**

36 bits (NChar = K/D bit + 8 bits data)

Fig. 3.19 Broadcast message format, composed by 4 SpaceFibre words

A common problem in networking is how to avoid to receive multiple copies of a broadcasted packet or making packets to circulate forever. In Ethernet, this problem is solved through well-known algorithms such as Spanning Tree Protocol [13] and Shortest Path Bridging [13]. These protocols try to create loop-free paths in the network, with every node sharing information about its own links to every other node in the network (link-state algorithms). This requires a preliminary phase (which converges after some time) and constant traffic to adapt to possible changes in the network. SpaceFibre adopts a different approach, with an ad-hoc algorithm to prevent Broadcast message flooding that does not require a set-up phase or the exchanging of special messages. Every Routing Switch keeps track of the following information for each Broadcast Channel, i.e. for each possible source of Broadcast messages in the network:

- is_first[chnl]: it identifies whether this is the first broadcast following a reset of the device for chnl;
- latest_arrival_port[chnl]: for each channel, the Routing Switch stores the index of the port where it has received the latest broadcast;
- timeout[chnl]: a timeout is associated with each channel. The timer starts to count when a broadcast is accepted and forwarded. According to the standard, the timeout must be set as slightly longer than the time required for a broadcast message to travel through the longest path of the network when all links in the network are operating.

SpaceFibre foresees that, when a new Broadcast message is received by a Routing Switch, it is accepted and thus forwarded to every other port except the input port if one or multiple of the following conditions hold:

- is_first[chnl] = 1;
- the arrival port is equal to latest_arrival_port[chnl]. This clause allows to accept messages coming from the same link while discarding the ones coming from other links, that are most likely duplicates;
- the timeout[chnl] has expired and the LATE flag of the broadcast is not set. This condition causes the Routing Switch to accept broadcasts from a different port but only after a maximum timeout and if the broadcast is not too old, handling the case in which the previous path is broken.

Consider that this mechanism could bring to the loss of some Broadcast messages, in the time-window while a timeout is not expired yet but some links are broken.

3.3.1.7 Wormhole Timeout

The Routing Switch here presented has two additional features concerning the SpaceFibre standard: the Wormhole Timeout and the Timeslot Guardian. The need for them arose in the study of the network-level FDIR mechanisms to implement in a SpaceFibre network presented in [19, 20]. The problem is that, while VNs are separated among each other and virtually independent thanks to the QoS mechanism, they also come in limited number: a SpaceFibre port can support a number of VNs equal to the number of VCs it has. This means that, in the very best case, a SpaceFibre port supports 32 separate VNs, i.e. 32 independent channels. More realistically, however, a SpaceFibre port will support 2, 4 or 8 VCs. If the number of source nodes is high, it is obvious that it is impossible to put each source node in a different VN, thus there must be several sources that compete for the same VC. This situation can lead to a problem in case the source node that took the right to transmit (it has the wormhole) is faulty and generate an infinite packet, without never releasing the wormhole. In this case, the VC Timeout specified in the standard is of no-help, because NChars are transmitted from the input to the output port, hence the timeout will never expire. To overcome this situation, a Wormhole Timeout mechanism has been proposed. The Wormhole Timeout defines the maximum time a wormhole can last, independently from the actual passage of NChars. In this Routing Switch, a different timeout can be specified for each <Port, VC> and it can be enabled or disabled by the user. Note that to specify the value of a Wormhole Timeout, it is necessary to know the maximum size of the packets that will be transmitted on that VC to estimate the time they will take to be transmitted. This assumption is realistic on real satellite networks, even if the SpaceWire and SpaceFibre standards allow to have infinite packets. Figure 3.20 shows an example where two nodes, one sane (blue) and one faulty (red) compete for the same output <Port, VC> in the Routing Switch. The sane node can transmit its packet within the Wormhole Timeout time, so everything goes fine. The faulty node however tries to send an infinite packet (or longer than expected). When the timeout expires, the wormhole open by the faulty node is closed by the Routing Switch and an End Error Packet (EEP) is appended to the transmitted portion of the packet, while the remaining is spilled. This mechanism, if used together with a Round Robin scheduling policy, allows calculating the worst-case blocking time for

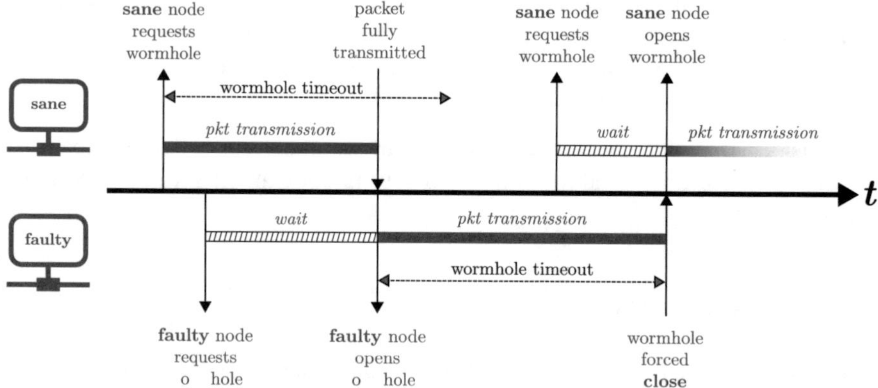

Fig. 3.20 The Wormhole Timeout prevents a the faulty node to take the VC forever

source nodes even in case of faulty nodes sharing the same VN. In particular:

$$MaximumBlockingTime_j = \sum_{i=0}^{allnodes} WormholeTimeout_i - WormholeTimeout_j \tag{3.1}$$

This is an important result because, in critical networks, it is necessary to have a maximum guaranteed latency. Again, the Wormhole Timeout and VC Timeout are two different mechanisms preventing two different error conditions:

- the Wormhole Timeout prevents faulty source nodes from jamming the network;
- the VC Timeout prevents faulty receiver nodes from blocking the reception (especially in the case of Multicast).

3.3.1.8 Timeslot Guardian

The other mechanism studied to implement network-level FDIR is the Timeslot Guardian. In a SpaceFibre port, each VC can transmit only in its assigned timeslots, where the total number of a timeslot is 64. However, when the number of VCs is low, it can be useful to partition the timeslots even inside the same VC. As explained in the previous section, when multiple sources share the same output VC, one faulty node can prevent the other from transmitting. The proposed solution is to use a Timeslot Guardian in each port of the Routing Switch. The Timeslot Guardian must be set with the same configuration of the far-end port to which it is connected and it checks whether the receiving data stream happens in the allowed timeslots. In case the wormhole open by a port continues in a non-allowed timeslot, the wormhole is immediately closed and an EEP is appended. In Fig. 3.21, Node

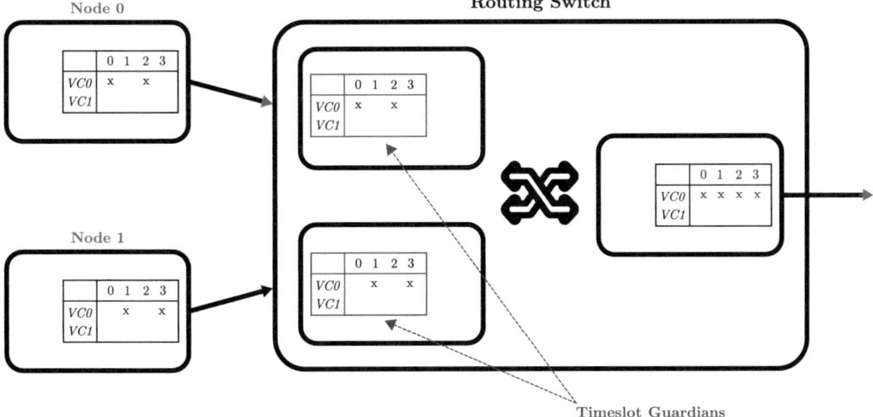

Fig. 3.21 Routing Switch implementing the Timeslot Guardians in the two input ports

0 and Node 1 share the same output VC in the output port of the Routing Switch but they are scheduled to transmit in different timeslots. According to the protocol, when the timeslot changes, the source VC stops to transmit but the wormhole stays open, preventing the other source node to get permission to transmit. The Timeslot Guardian checks that the packet is finished within the allowed timeslot, and closes the wormhole otherwise. As for the Wormhole Timeout, also this mechanism allows to calculate the maximum blocking time for a node even when it shares the same VNs with other faulty nodes.

In particular, a sane node always experiences zero blocking time: as soon as it is its timeslot to transmit, every other wormhole from every other node is immediately closed in the Routing Switch.

3.3.2 SpaceFibre Full Router

In the previous section, the core of the SpaceFibre Routing Switch is described. However, a fully functional router needs a lot of additional logic to be configured and used in a real-case scenario. As demonstrated by successful commercial products such as the Gaisler GR718B [11] and the STAR-Dundee SpaceWire Router 10X [24], the RMAP protocol is the standard way to access the configuration space of the router. Moreover, a hot topic in the development of the SpaceFibre standard is its compatibility with SpaceWire at the Network Level, which means the existence of a router able to process both kinds of protocols and bridge them together. These two needs led to the extension of the Routing Switch described in the section above to implement a full-fledged and ready-to-use router. An overview of the Full Router architecture developed in this work is shown in Fig. 3.22. There are several things to note about the architecture of the Full Router:

- the Routing Switch does not have modification concerning the one described above, meaning that it is SpaceFibre-compliant only. A SpaceWire-to-SpaceFibre Adapter is used to bridge one (or more) of its port connections to a SpaceWire port;
- the internal port with index 0 is not connected to a physical port but it is used as Configuration Port. This means that, as stated in the two standards, it must be the destination of configuration packets;
- a Register File contains the status and configuration parameters for the Routing Switch and all the ports (Configuration, SpaceWire and SpaceFibre). The Register File is organised according to a certain configuration space described in the next section, which defines the addresses of the registers;
- the Register File is read and written by an RMAP Target engine, which accesses it depending on the RMAP commands received through the Configuration Port.

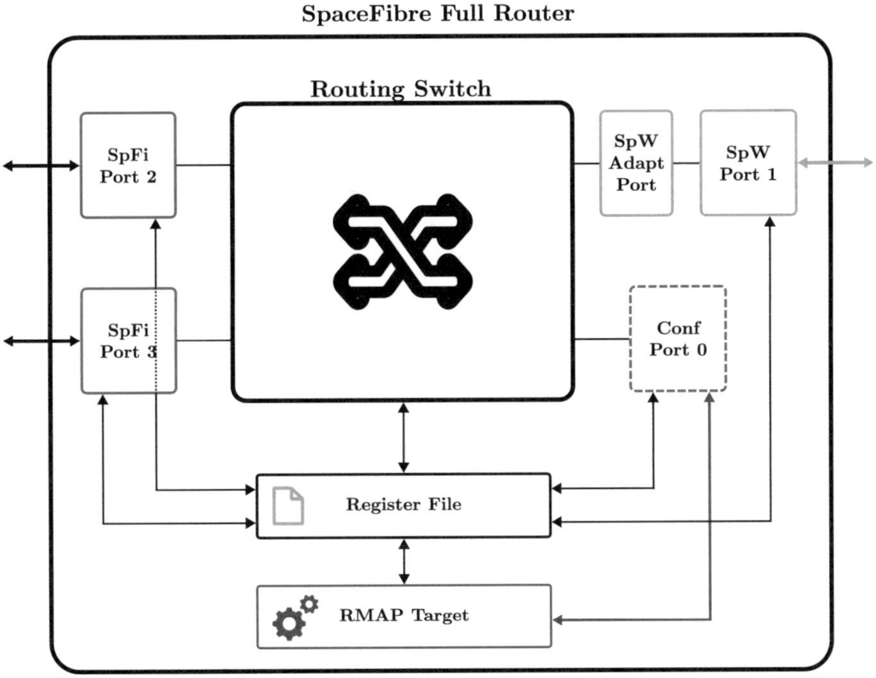

Fig. 3.22 SpaceFibre Full Router architecture overview

3.3.2.1 Configuration Space

A SpaceFibre Router is a complex system on its own, with many dynamically configurable parameters. The most obvious examples are the Routing Table and the Virtual Network Table. However, in addition to the router parameters, also its SpaceFibre (and SpaceWire) ports must be dynamically configurable. To this goal, a well-organised configuration space has to be defined. The definition of the configuration space has been a problem in the past, with different vendors creating their own non-interoperable devices. A possible solution has been proposed with the definition of the Network Discovery and Configuration Protocol (NDCP) [30] and its associated SpaceMAN protocol [17]. The approach proposed in NDCP is to overcome the problem of different configuration spaces among vendors by identifying the configuration and status registers not through their memory address (implementation dependent) but common labels (implementation-independent). However, NDCP has not been implemented in any commercial product yet and it is limited to SpaceWire, with no support for SpaceFibre. Given that SpaceFibre is still under the standardisation process, an open configuration space has been proposed in [21]. The SpaceFibre router here presented follow this configuration space scheme, which is summarised in Fig. 3.23.

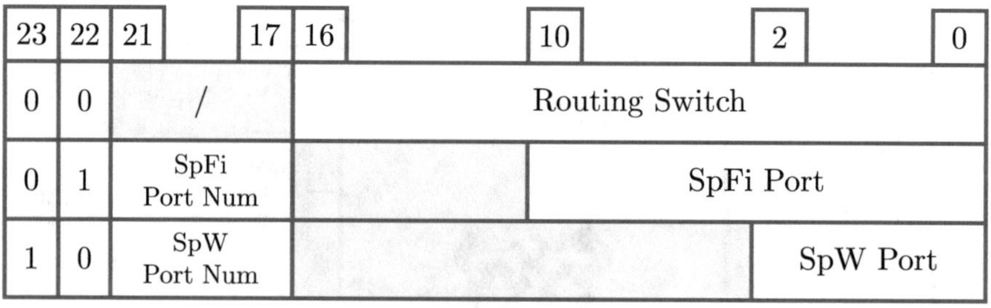

23	22	21 … 17	16 …	10 …	2 … 0
0	0	/	Routing Switch		
0	1	SpFi Port Num		SpFi Port	
1	0	SpW Port Num			SpW Port

Fig. 3.23 Higher separation in the Full Router configuration space

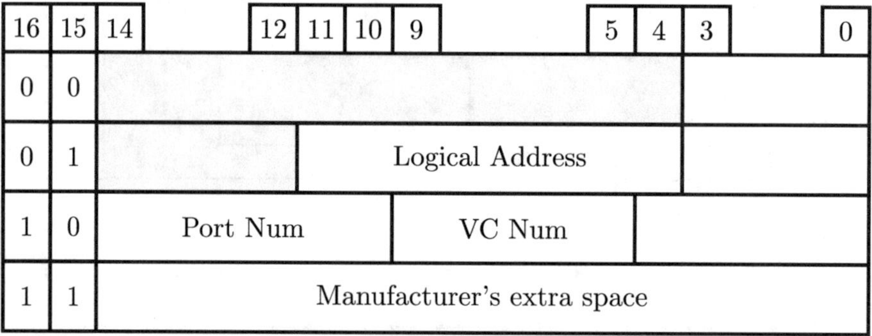

Fig. 3.24 Subset of the configuration space comprising the configuration and status parameters of the Routing Switch

As can be seen, the configuration space is not contiguous concerning the addressing, meaning that there are "holes". This leads to a waste of addresses but simplifies a lot the implementation of the logic managing the register file. The waste of addresses is not seen as a problem given that the entire configuration space is addressed anyway with only 24 bits and most systems allow 32 bits addressing (RMAP protocol supports up to 40 bits addressing). The configuration space depicted in Fig. 3.23 is divided into three main memory regions:

- the Routing Switch configuration space (bits[23:22] = "00");
- the SpaceFibre ports configuration space (bits[23:22] = "01"). Because a router contains multiple SpaceFibre ports, the address bits [21:17] are used to discriminate the specific port referred;
- the SpaceWire ports configuration space (bits [23:22] = "10"). As it happens for SpaceFibre, the bits [21:17] are used to discriminate the SpaceWire port among the ones in the router.

Note that, according to both standards, the maximum number of ports in a router is 32, thus the 5 bits to index the port. Figure 3.24 depicts the configuration space of the Routing Switch (the "expansion" of the first row of Fig. 3.23). As can be seen, it is divided into four regions:

- a generic configuration region (bits [16:15] = "00"), where the global parameters of the Routing Switch can be read and written. Examples of this are the Product_ID and all the parameters related to the Broadcast Service;
- the Routing Table region (bits [16:15] = "01"). The Routing Table contains, for each Logical Address, the index of the physical output port associated with it. Given that there are 224 possible Logical Addresses, the eight bits [11:4] are used to access this table. For each Logical Address, 9 bytes can be configured, addressed through the bits [3:0]:
  - bytes [0:3] are used to specify the output ports as a 32-bit register;
  - bytes [4:7] are used to specify whether the header must be removed or not when forwarding the packet to the correspondent output port;
  - byte 8 is used as a flag to discriminate between Multicast or normal/Group Adaptive routing;
- the Virtual Network Table region (bits [16:15] = "10"). The Virtual Network Table has an entry for each tuple <Port, VC>, identified by the bits [14:5]. Each entry of the Virtual Network Table contains the VN to which <Port, VC> belongs to plus additional debug information, such as decode errors or expired timeouts.
- at the bottom of this space, a manufacturer's extra configuration space is left (bits [16:15] = "11"). This is useful to allow each manufacturer to implement his specific features into the Routing Switch while maintaining compatibility with other instruments, keeping a "standard" configuration space.

The configuration space for the SpaceFibre and SpaceWire ports are not reported here because their organisation reflects the configuration parameters described in the two standards.

3.3.2.2 RMAP Target Engine

The Register File is accessed through an RMAP Target engine that has been developed specifically for the SpaceFibre Full Router. To develop a generic RMAP engine that can be used also in other projects, it takes as input and produces as output bytes instead of words, with an ad-hoc Word-to-Byte Translator module realising the width conversion. This also allows using words of configurable width (via a VHDL Generic) without modifying the internal logic. The RMAP Target engine fully implements the RMAP standard [7]. In particular, it implements:

- acknowledgement support: write commands can require to send an acknowledgement to the source of the RMAP request containing the status result of the operation;
- verification support: write commands can store their payload in a Verification Buffer before actually writing into the Register File. The write operation takes

place only if the CRC of the entire payload is approved. Of course, the size of the Verification Buffer represents an upper limit for the size of the write commands using the verification feature;

- incremental support: both read and write commands can be executed with an incremental or fixed address.

The RMAP Target engine supports also the Extended Memory Address feature, e.g. it can process addresses up to 48 bits, although the Register File uses only the first 24 bits and ignores the others.

3.3.2.3 SpaceWire Adapter Port

The last module implemented for the SpaceFibre Full Router is the SpaceWire Adapter Port, which can be used to plug a SpaceWire CoDec into the Routing Switch without any modification to their source code. The principle adopted to implement the Adapter Port is the same used in the Simulator for HIgh-speed Network (SHINe) [26]. The Adapter Port must: (1) translate SpaceFibre to SpaceWire packets and vice versa and (2) translate Broadcast messages to TimeCodes and vice versa. Figure 3.25 shows an overview of the SpaceWire Adapter Port. Its interface towards the Routing Switch must be equal to a SpaceFibre port interface, hence it has a packet interface transferring words (with a width specified via a VHDL generic) and a Broadcast message interface. On the other side, towards the SpaceWire port, it has a packet interface transferring bytes (plus the control bit, for EOP and EEP insertion) and a TimeCode interface. More in detail, the Adapter Port implements these two functions:

- Packet Translator: this block is responsible for translating the SpaceFibre packets into SpaceWire packets. To this goal, it must:

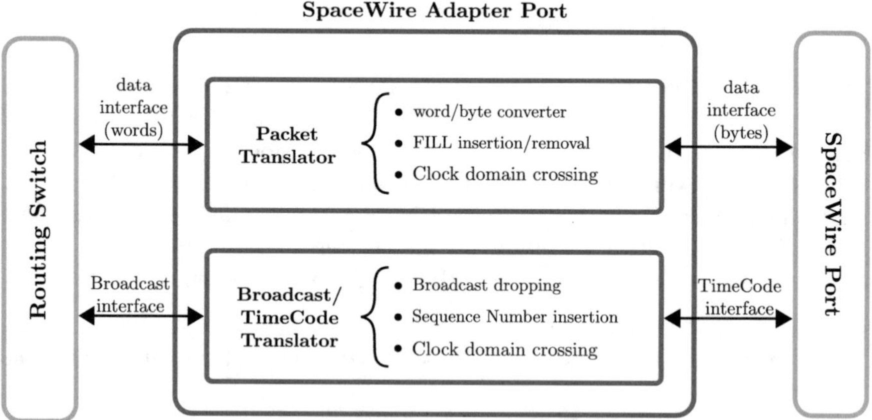

Fig. 3.25 SpaceWire Adapter Port interfacing the Routing Switch to a SpaceWire port

  - convert words (36, 72 or 144 bits wide) into a byte stream (9 bits wide) and vice versa. This is implemented with a simple FSM;
  - remove possible SpaceFibre FILL characters when transferring from the Routing Switch to the SpaceWire port or insert them in the other case. FILL characters are used by SpaceFibre to pad the packet payload to the word width and are not needed (and cannot be processed) by SpaceWire;
  - implement the clock domain crossing between the system clock used by the Routing Switch and the SpaceWire port clock. This is realised through an asynchronous FIFO;

- Broadcast/TimeCode Translator: this block realises the translation between the Broadcast messages used by the Routing Switch and the TimeCodes of the SpaceWire port. Its roles are:
  - to implement the possible Broadcast dropping. Broadcast messages can carry a wide range of different message types and only one corresponds to the SpaceWire TimeCodes. Broadcast messages with a type different from TimeCode cannot be forwarded into a SpaceWire network, thus must be dropped;
  - when trying to forward a TimeCode into a SpaceFibre network, the TimeCode must be transformed into a Broadcast message of a specific type and with a specific Sequence Number. According to the SpaceFibre standard, the Broadcast Sequence Number uniquely identifies the source node generating the message and it is used by the loop prevention mechanism. In this case, the Adapter Port creates a Broadcast message carrying the TimeCode and tags it with a Sequence Number that can be specified via a VHDL generic. This means that the Routing Switch acts as the creator of the Broadcast message, hence the SpaceFibre network acts as a backbone for SpaceWire TimeCodes;
  - also, in this case, the clock domain crossing must be implemented. A small asynchronous FIFO is used.

To summarise, the SpaceWire Adapter Port is a small component allowing to bridge of the SpaceWire protocol with the SpaceFibre protocol used by the Routing Switch. This allows a seamless interconnection between the two kinds of network.

3.4 SpaceART®: A Complete Mixed SpaceWire/SpaceFibre EGSE

3.4.1 EGSE State of the Art

Test equipment devices play a key role during the whole space mission life-cycle. An example of a typical project life-cycle is shown in Fig. 3.26. Within a typical project life-cycle (Fig. 3.26), the use of Electrical Ground Support Equipment (EGSE) is crucial. After the initial stages (Phases 0 *Mission analysis/needs iden-*

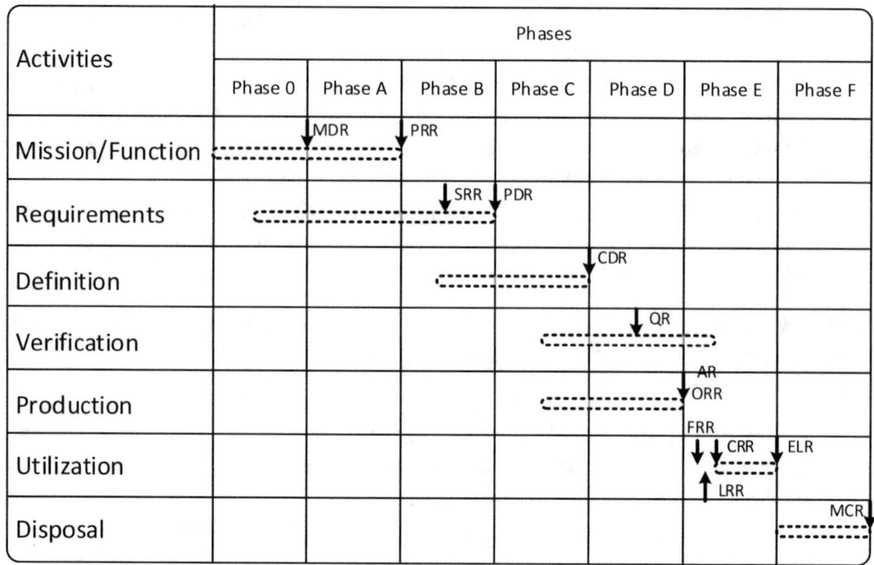

Activities
Phases
Phase 0
Phase A
Phase B
Phase C
Phase D
Phase E
Phase F
Mission/Function
MDR
PRR
Requirements
SRR
PDR
Definition
CDR
Verification
QR
Production
AR
ORR
FRR
Utilization
CRR
ELR
LRR
Disposal
MCR

Fig. 3.26 Typical project life-cycle [8]

tification, Phase A *Feasibility* and Phase B *Preliminary Definition*) where all project activities are mainly focused on requirements definitions, activities and resources identification and pre-development tasks; satellites designers need to have available EGSE systems during the Phase C (*detailed definition*) and Phase D *qualification and production*. During these phases, all the activities needed to develop and qualify space and ground segments are performed before the launch and utilisation of spacecraft (Phase E -*Utilisation*). [8]

EGSE are tools used by satellite and sub-system integrators to test and validate the electrical, functional and performance of the satellite before its launch into space. Providing the necessary hardware and software elements, the EGSE system enables spacecraft manufacturers to perform satellite tests by emulating the entire part of the missing sub-system to assure full compatibility once integrated within the overall payload. Figure 3.27 shows a basic block diagram of EGSE system architecture. Such test equipment is composed of several interfaces to communicate with a DUT and they can generate stimuli in compliance with DUT requirements analysing its output and verifying the correct behaviours. Moreover, EGSE offers host-PC interfaces to enable the end-user to configure and execute his test chain.

As already mentioned, EGSE can emulate missing components of space system by simulating their behaviour through compatible interfaces. However, device emulation is not the only use of the EGSE system. They may test the compliance with the reference standard of DUT (Protocol Conformance Tester) or can also analyse the traffic between two nodes in an unobtrusive way (Sniffer). Within the process of SpaceFibre standard adoption, it is clear that a set of SpaceFibre compatible EGSE systems should be made available to satellite integrators. Fol-

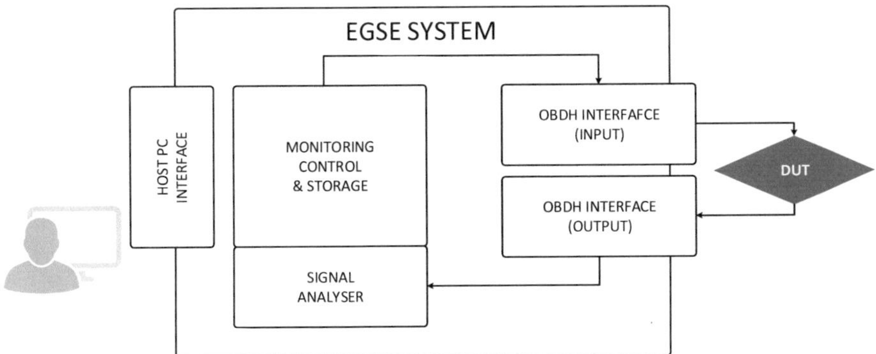

Fig. 3.27 EGSE system—general view

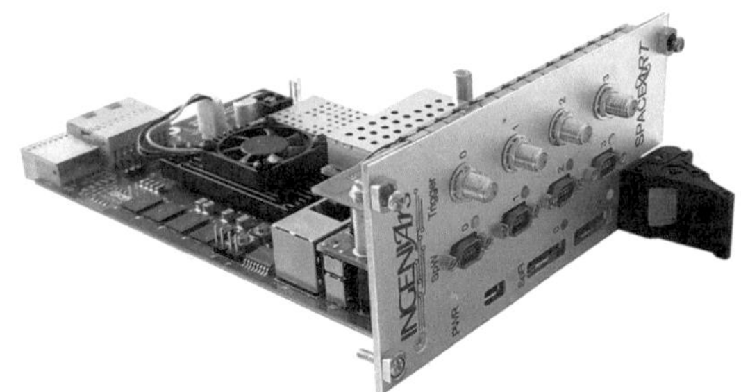

Fig. 3.28 SpaceART EGSE solution—cPCI Board

lowing the SpaceFibre/SpaceWire Analyser Real-Time (SpaceART®) is presented. SpaceART®is one of the first SpaceFibre compatible EGSE systems. After a brief description of SpaceART®features, a high-level overview of the most important building blocks needed to implement an EGSE system is given.

3.4.2 SpaceART®Main Features

SpaceART®EGSE system is suitable to support the development of sub-modules based on state-of-the-art space communication standards defined by the ESA: SpaceWire and SpaceFibre. Figure 3.28 shows a picture of SpaceART®board.

SpaceART®can be used as a key component of the EGSE system to provide SpaceWire/SpaceFibre device emulation and/or link analysis in one single and practical solution.

From the SpaceFibre side, SpaceART®EGSE enables the end-user to transmit and to receive SpaceFibre data through 8 different VCs among each of Space-Fibre interfaces. SpaceFibre data coming from any VC can be easily re-routed through SpaceWire ports thanks to the SpaceFibre/SpaceWire bridging capabilities exploited from SpaceART®firmware. On the SpaceWire communication side

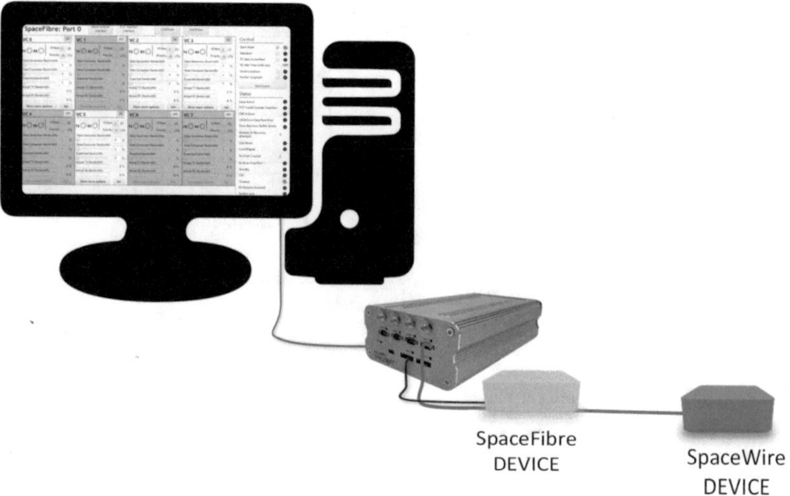

Fig. 3.29 SpaceART®SpW/SpFi bridging typical scenario

instead, SpaceART®offers a configurable SpW transmit speed from 10 Mbps up to 400 Mbps for each of the 4 available SpaceWire interfaces. As per the above-mentioned SpaceFibre interfaces, the SpaceWire data received from one of the SpaceWire interfaces can be re-routed and retransmitted through any SpaceFibre VC exploiting the SpaceWire/SpaceFibre bridging feature (Fig. 3.29). Furthermore, SpaceART®is equipped with a set of features implemented to help the end-user during the integration tests; trace memory and error injection capabilities are two examples of SpaceART®embedded debug features. One of the strongest points of SpaceART®is the capability to generate and to consume SpaceFibre and/or SpaceWire packet exploiting the PCIe communication. Real-time of SpaceFibre and/or SpaceWire packets record and playback is a feature which makes it possible to perform pre-scheduled test scenario directly from the Host-PC by keeping the EGSE system unobtrusive in terms of bandwidth saturation. Following all the SpaceART®accessible interfaces are briefly reported:

Accessible interfaces:

- 4 standard 9-way Micro-miniature D-type SpaceWire socket connectors:
- 2 eSATA interfaces configurable as:
  - SpaceFibre interfaces running up to 3125 Gbps;
  - WizardLink compatible interfaces;
- 4 SMA connectors programmable as trigger input/output;
- Ethernet and compact Peripheral Component Interconnect express (cPCIe) interfaces for host-PC communication;
- USB 2.0 interface for system upgrade.

3.4.3 SpaceART®Architecture Overview

SpaceART®is an embedded system based on System on Chip (SoC) FPGA device which integrates a microprocessor, different interfaces, memory and other subsystems. Using an SoC FPGA allows you to leverage both the speed of physical devices such as microcontrollers and peripherals as well as memory controllers and high-speed communication interfaces. Also, all the capabilities of the SoC are seamlessly interfaced to the Programmable Logic (PL) of the FPGA making it easy to integrate and interface custom architectures directly with the microcontroller. The use of a microprocessor gives the system a high level of flexibility and the possibility to quickly reprogram the firmware to interact with the hardware. The microprocessor is also helpful for interfacing SpaceART®with the host-PC; it handles both, low-level and high-level communications. Low-level communication is related to on-board communication between peripherals and microprocessor. The high-level communication instead is related to the communication between the microprocessor and the host-PC; it can be performed through different high-speed protocols such as Gigabit Ethernet and/or PCI express (PCIe). In Fig. 3.30 the overall architecture of the SpaceART®EGSE system is shown.

As depicted in the figure above, the general architecture can be sectioned into four macro-areas:

- **Microcontroller and Memory**: the microcontroller is responsible for the configuration and settings of the entire system. It also handles the command (CMD) & control (CTRL) mechanism and the data stream from/to serial-interfaces to/from the host-PC.
- **Serial protocol interfaces**: this section includes all the IPs responsible to implement the serial communication protocol compliant to the standard. Each serial protocol IPs is equipped with: a physical connector cable for communication with the DUT and with both slave and master bus interfaces for communication with the microcontroller.

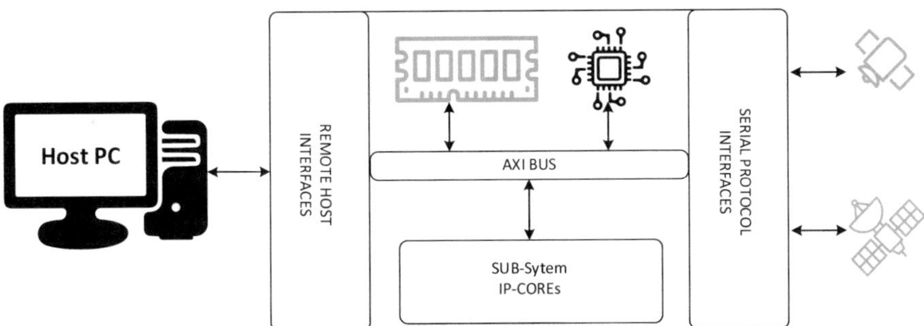

Fig. 3.30 Overall architecture of SpaceART®EGSE

- **SUB-System IP-COREs**: this macro-section comprises a set of auxiliary modules developed to implement additional functionalities necessary for EGSE purposes.
- **Remote host interfaces**: as the name suggests, this area deals with the host-PC communication. In this section, the protocol used for the host-PC communication is implemented.

The target device is the Xilinx Zynq-7000 FPGA. The Zynq FPGA provides SpaceART ®of an integrated hard-core microprocessor, memory interfaces and many on-board peripheral.

Figure 3.31 shows the functional block diagram of the SpaceART ®architecture implemented into the Xilinx Zynq-7000 FPGA target device.

The SpaceART®is developed as an ARM-based embedded system; it uses the AMBA AXI4 high-speed bus to interconnect peripherals. AXI4 is the bus specification for high-performance systems and it allows achieving Real-Time

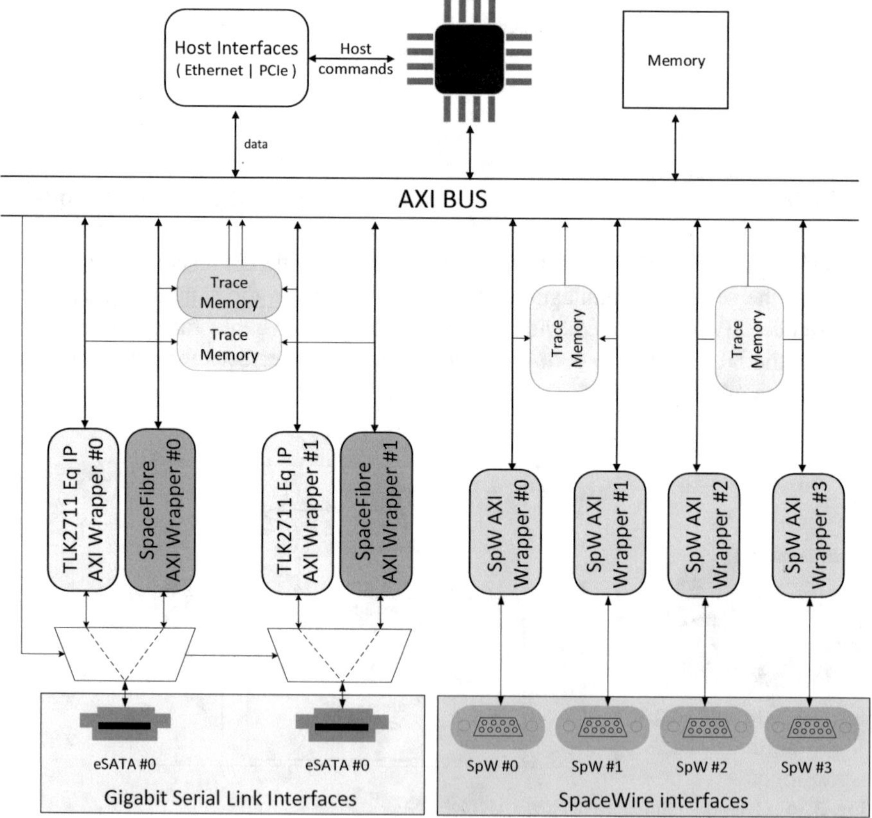

Fig. 3.31 SpaceART®system functional block diagram [22]

data transfer among SpaceART®architecture. SpFi/SpW data can be streamed from/to host-PC by PCIe or Ethernet communication protocols. Both high—SpaceFibre/WizardLink—and medium—SpaceWire—speed communication protocol IPs are described in Very high-speed integrated circuits Hardware Description Language (VHDL) and their hardware architecture they also include DMA engines, hardware packet generator/consumer and error injection modules. The combination of hard-core processor and hardware developed IPs allows having at the same time a high software execution speed and the handling of all the peripherals in the system. The whole system is partitioned into hardware (HW) and software (SW) domains. Thanks to the hardware/software partitioning are possible to upgrade the system by just reprogramming the microprocessor in C/C++ language without modifying the hardware. The hardware is designed for the different specific functionalities/peripherals, and each sub-system includes configuration registers that can be read and written by the on-board microprocessor. The employment of dedicated hardware modules gives the overall architecture advantages in terms of speed, resources and area-saving and reduction of power consumption.

Besides the architectural part that makes up an EGSE, another very important aspect that concerns these systems is the firmware that runs on them. To fully exploit the peculiarities of the architectures that constitute this type of systems, the firmware must be written in such a way as not to act as a bottleneck in terms of data transfer speed from the host to the interfaces and vice versa.

The Firmware Architecture of the SpaceART®EGSE system is based on the Petalinux Operative System (OS) [33]. Petalinux is a Xilinx customised distribution of Linux Kernel which makes it possible to use all the main Linux functionalities keeping low the impact of memory usage and performance overhead. Exploiting the Linux ecosystem features it is possible not only to take full advantages of both dual-core ARM available on SoC FPGA, but it is also possible to make uses of POSIX Threads (pthread) library for implement well-structured applications [18].

The end-user can interact with SpaceART®EGSE through the SpaceART® software application. SpaceART®can be programmed by using both APIs and GUI. APIs have been developed to leave to end-users the possibility to write their test chain and to execute it from the host-PC. While APIs are preferable for building custom test chains to be run directly from the host via scripts, in some cases it is essential to use a GUI that helps the user to easily identify corner case scenarios. Being able to have on-screen the output of the data streaming when a certain condition occurs is an example of use for which the GUI becomes a very useful tool but it is also important for a simple configuration of interfaces to verify the correct link establishment before start with a complex test chain. The SpaceART®EGSE solution is also equipped with a general-purpose GUI which enable the user to easily interact with the instrument without any programmable language knowledge.

To control SpaceART®from host-PC performing custom test chain, SpaceART® is equipped with C, C++, Java API libraries. C APIs offer a simple, yet powerful interface to use all SpaceART®functionalities. Notice that API functionalities follow the Client-Server approach, in which SpaceART®acts as a server receiving data and commands, and returning data and status information on host request

(Fig. 3.32). All request operations can be performed in both blocking and non-blocking mode.

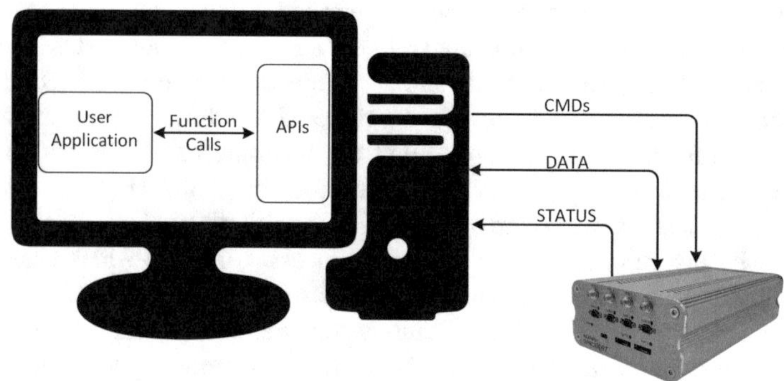

Fig. 3.32 Software architecture block diagram

References

1. Agarwal, R., Solanki, G. S., & Sharma, S. (2013). Power optimization of high speed pipelined 8b/10b encoder. *International Journal of Innovative Technology & Exploring Engineering, 3*(7). ISSN: 2278-3075.
2. Cummings, C. E. (2002). Simulation and synthesis techniques for asynchronous FIFO design. In *SNUG 2002 (Synopsys Users Group Conference, San Jose, CA, 2002) User Papers*, 2002.
3. Chiabrera, M. (2012). *Apparatus and Method for Transmitting and Recovering Multi-Lane Encoded Data Streams Using a Reduced Number of Lanes*, September 4, 2012. US Patent 8,259,760.
4. Dinelli, G., Marino, A., Dello Sterpaio, L., Leoni, A., Fanucci, L., Nannipieri, P., & Davalle, D. (2020). A serial high-speed satellite communication codec: Design and implementation of a SpaceFibre interface. *Acta Astronautica, 169*, 206–215.
5. Dinelli, G., Nannipieri, P., & Fanucci, L. (2019). A configurable hardware word re-ordering block for multi-lane communication protocols: Design and use case. *IEICE Transactions on Fundamentals of Electronics, Communications and Computer Sciences, E102A*(5), 747–749.
6. Dinelli, G., Nannipieri, P., Marino, A., Fanucci, L., & Sterpaio, L. D. (2020). The very high-speed SpaceFibre multi-lane codec: Implementation and experimental performance evaluation. *Acta Astronautica, 179*, 462–470.
7. ESA Std. (2010). Ecss-e-st-50-52c. *SpaceWire-Remote Memory Access Protocol.*
8. ESA-ESTEC Std. (2009). Ecss-m-st-10c rev.1. *Space Project Management - Project Planning and Implementation.*
9. European Cooperation for Space Standardisation (2008). *Space product assurance – ASIC and FPGA development, ECSS-E-ST-60-02C*. European Cooperation for Space Standardisation, 2008.
10. European Cooperation for Space Standardisation (2019). *SpaceFibre – Very high-speed serial link, ECSS-E-ST-50-11C*. European Cooperation for Space Standardisation, 2019.

11. Habinc, S., Johansson, F., Hernandez, F., Sturesson, F., Siege, F., & Suess, M. (2016). Radiation-tolerant 18x SpaceWire router design and qualification for space application – GR718B: components, long paper. In *2016 International SpaceWire Conference (SpaceWire)* (pp. 1–6).
12. Hua, S.-L., Wang, Q., & Wang, D.-H. (2010). A 1.1 GHz 8b/10b encoder and decoder design. In *2010 Asia Pacific Conference on Postgraduate Research in Microelectronics and Electronics (PrimeAsia)* (pp. 138–141).
13. *IEEE 802.1Q-2018 - IEEE Standard for Local and Metropolitan Area Networks—Bridges and Bridged Networks* (2018). New York: IEEE.
14. Koopman, P., & Chakravarty, T. (2004). Cyclic redundancy code (CRC) polynomial selection for embedded networks. In *International Conference on Dependable Systems and Networks*, 2004 (pp. 145–154). New York: IEEE.
15. Korobkov, I. L. (2019). Algorithm for schedule-table's designing for SpaceFiber network technology. In *2018 Wave Electronics and its Application in Information and Telecommunication Systems (WECONF)*. New York: IEEE.
16. Kumar, N., Madan, R., & Deb, S. (2015). Pragmatic approaches to implement self-checking mechanism in UVM based testbench. In *2015 International Conference on Advances in Computer Engineering and Applications* (pp. 632–636). New York: IEEE.
17. Lancmański, P., Romanowski, K., Kollias, V. D., Hołubowicz, W., & Pogkas, N. (2014). Spaceman: A SpaceWire network management tool. In *2014 International SpaceWire Conference (SpaceWire)* (pp. 1–4).
18. Lawrence Livermore National Laboratory Blaise Barney (2015). *POSIX threads programming*. Available online: https://computing.llnl.gov/tutorials/pthreads/
19. Leoni, A. (2017). *An Example of Use of Union Generic Spec for SpaceFibre*.
20. Leoni, A., & Siegle, F. (2017). *Possible Requirements for the SpaceFibre Network Layer*.
21. Leoni, A., & Siegle, F. (2018). Standardization efforts for a network management and discovery protocol for SpaceFibre. In *SpaceWire-2018. Proceedings of the 8th International SpaceWire Conference*, 2018.
22. Leoni, A., Dello Sterpaio, L., Nannipieri, P., Dinelli, G., Benelli, G., Davalle, D., Marino, A., & Fanucci, L. (2020). SpaceArt SpaceWire and SpaceFibre analyser real-time. In *2020 IEEE 7th International Workshop on Metrology for AeroSpace (MetroAeroSpace)* (pp. 244–248). New York: IEEE.
23. Mastronarde, J. B. (2003). *Dual Input Lane Reordering Data Buffer*, January 21 2003. US Patent 6,510,472.
24. McClements, C., Kempf, G., Fischer, S., Fabry, P., Parkes, S., & Leon, A. (2007). SpaceWire router ASIC. In *2007 International SpaceWire Conference (SpaceWire)*.
25. Nannipieri, P., Davalle, D., & Fanucci, L. (2018). A novel parallel 8b/10b encoder: Architecture and comparison with classical solution. *IEICE Transactions on Fundamentals of Electronics, Communications and Computer Sciences, E101A*(7), 1120–1122.
26. Nannipieri, P., Davalle, D., Fanucci, L., Leoni, A., & Jameux, D. (2019). Shine: Simulator for satellite on-board high-speed networks featuring SpaceFibre and SpaceWire protocols. *Aerospace, 6*(4), 43.
27. Nannipieri, P., Leoni, A., & Fanucci, L. (2019). VHDL design of a SpaceFibre routing switch. *IEICE Transactions on Fundamentals of Electronics, Communications and Computer Sciences, E102A*(5), 729–731.
28. Parkes, S., McClements, C., McLaren, D., Florit, A. F., & Villafranca, A. G. (2015). SpaceFibre: A multi-gigabit/s interconnect for spacecraft onboard data handling. In *2015 IEEE Aerospace Conference*, June (Vol. 2015).
29. Qing-Sheng, H., Yu-Yun, S., & Liming, H. (2010). A 0.18 μm pipelined 8b10b encoder for a high-speed SerDes. In 2010 IEEE 12th International Conference on Communication Technology (pp. 1039–1042).

30. Tyczka, P., Hołubowicz, W., Renk, R., Kollias, V. D., Pogkas, N., Romanowski, K., & Jameux, D. (2016). SpaceWire network management using network discovery and configuration protocol: SpaceWire networks and protocols, short paper. In *2016 International SpaceWire Conference (SpaceWire)* (pp. 1–7).
31. Widmer, A. X., & Franaszek, P. A. (1983). A dc-balanced, partitioned-block, 8b/10b transmission code. *IBM Journal of Research and Development, 27*(5), 440–451 (1983)
32. Woodcock, J. , Larsen, P.G., Bicarregui, J., & Fitzgerald, J. (2009). Formal methods: Practice and experience. *ACM Computing Surveys, 41*(4), 1–36. Cited by 349.
33. Xilinx (2016). *Petalinux Tools Documentation, Reference Guide UG1144 (v2016.3).*

Chapter 4
Interoperability Test: How to Verify Compliance to the Standard

The maturity of a protocol is related to the interoperability between separately developed implementations. The SpaceFibre standard has a descriptive style that varies depending on the layer: there is some layer where the hardware implementation is straightforward from the requirements and leaves no space for interpretation, e.g. physical and lane layers. On the other hand, Data-Link and Multi-Lane layers in some passages describe the protocol at a very high level, leaving space for interpretation of requirements which can lead to significantly different hardware implementations. This aspect can potentially lead to interoperability issues. Moreover, the protocol standardisation is very recent (2019), and up to our knowledge, there is no hardware conformance tester available or planned to be developed. The market is currently too small to justify the development of such devices and the low number of solutions available does not push in that direction. Even without a dedicated conformance tester, it is however necessary to verify the interoperability of separately developed implementations of the standard: this needs to happen both in the interest of the single vendors and in the interest of the SpaceFibre community; having separately developed implementations fully interoperable contributes to confirming the high-level quality of the standard. In the following, we will propose a test plan to be followed to verify interoperability between two separately implemented CoDecs. We will also report on a performed test campaign, carried out with the CoDecs available on the market.

A simplified version of the proposed test set-up is shown in Fig. 4.1. Each device has been tested in loopback and connected with another device. In this phase, it is of great help to employ devices connected to link analyser tools, so that the test engineer can easily control via a dedicated interface on the host PC all the communication parameters and can provide all the necessary stimuli to the communication link.

P. Nannipieri et al., *Next-Generation High-Speed Satellite Interconnect*,
https://doi.org/10.1007/978-3-030-77044-0_4

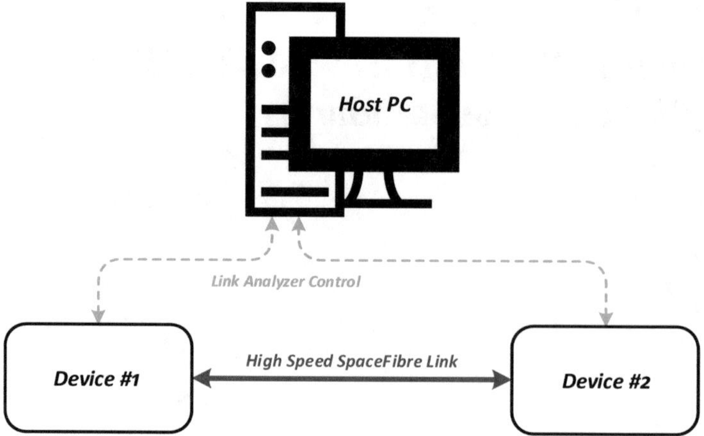

Fig. 4.1 Interoperability test set-up

4.1 Test Plan

The tests to be carried out to demonstrate the interoperability between different SpaceFibre IP cores needs to be derived directly from the standard specifications. In the following, we will present the description of a test plan that can be used to assess compliance and interoperability of separately developed SpaceFibre CoDecs. Please note that we are not presenting a conformance tester: our aim is not to verify one by one the compliance of a CoDec to specific requirements, as it is something that the CoDec developer has to do on its own during the verification stage. Instead, we aim to verify that the interpretation of those requirements is compatible with two separately developed interfaces. The proposed test plan is intended to be used as a reference for future system developers to assess device interoperability, to be mapped on the appropriate link analyser or CoDec configuration interface. The test was focused on the following areas and mechanisms of the communication protocol:

- **LL:** Lane layer initialisation state machine. It is the mechanism responsible for establishing the link, following a detailed handshake procedure, implemented through a Finite State Machine. It is one of the most critical components of the protocol: it needs to be robust against all the possible error situations, and it needs to be able to re-establish the connection between the two ends of the link.
- **ML:** Multi-Lane layer synchronisation state machine. It is the mechanism that reshuffles the data flow according to the real-time varying number of active lanes, without losing any data, taking also into account special lane configurations (hot redundant, Tx only and Rx only).
- **DL:** Data-Link layer compatibility. Different high-level mechanisms are provided by the Data-Link layer: data packets are encapsulated and CRC code is calculated and attached to every single frame to be able to detect possible error

situations: data may also be scrambled or not. Finally, a series of mechanisms shall regulate all retry procedures. Even though the standard is very clear about the related requirements, it is fundamental to check whether separate implementations are compatible with each other or not at this level.

- **QoS:** Quality of Service interpretation. Still, in the Data-Link layer, a series of services is provided to control the quality of the link: the link itself is indeed shared between separated virtual channels, which are scheduled to transmit over the same link on the base of various QoS parameters, such as bandwidth allocation, priority and timeslots allocation. It is necessary to verify that different QoS implementations are compatible with one another.

Table 4.1 briefly describes the hardware test plan. The presented test plan aims to stimulate all the core features of a SpaceFibre link. Some of the tests provide complex stimuli to the devices under test; for example, tests involving manual

Table 4.1 Interoperability test plan

Test ID	Test description
LL-1	Verify that the handshake procedure works smoothly (no NACKs)
LL-2	Verify that the handshake after the link/lane reset of both ends works smoothly
LL-3	Verify that the link is able to be established after manual disconnection of the cable (multiple reconnections/retries)
DL-1	Verify that full bandwidth communication occurs without any loss of data/nacks
DL-2	Verify that full bandwidth communication occurs without loss of data when injecting errors
DL-3	Communication with scrambling activated/deactivated
DL-4	Broadcast communication test
DL-5	Verify that no data loss occurs if the cable is manually disconnected during normal communication
ML-1	Verify that full bandwidth communication occurs exploiting all available lanes without any loss of data with a BER>0
ML-2	Verify that full bandwidth communication occurs exploiting all available lanes without any loss of data with real-time lane(s) cable(s) connection/disconnection
ML-3	Verify that, in the presence of a hot redundant lane, it takes the place of a disconnected communication lane, without any loss of data
ML-4	Verify that full bandwidth communication occurs exploiting all available lanes, including a Tx only lane
ML-5	Verify that full bandwidth communication occurs exploiting all available lanes, including a Rx only lane
QoS-1	Multiple VCs competing for the link
QoS-2	Full communication with small packet sizes (1–50)
QoS-3	Full communication with medium packet sizes (51–1000)
QoS-4	Full communication with big packet sizes (9000)
QoS-5	Verify that broadcast transmission does not exceed maximum allowed bandwidth value
QoS-6	Verify that timeslot setting is correctly implemented
QoS-7	Verify that priority setting is correctly implemented

disconnection of the eSATA cable during normal operations produce a high number of nested retries due to multiple and narrow link connections and disconnections, generated by connector bouncing. Considering that the objective of the test plan is to demonstrate the interoperability of the core features, all the core blocks of a SpaceFibre protocol shall be investigated, monitoring the link status and behaviour and in all the proposed corner case scenarios.

4.2 Test Campaign Results

In this section, we will present and report the results of a test campaign performed according to the previous test plan. This report provides details about the currently available CoDecs and, at the same time, gives a valid template for future interoperability tests of new CoDecs.

Keeping in mind the interoperability demonstration objective between different SpFi IPs, the latest available version of each single SpFi IP has been used. The mentioned IPs have been mapped onto various hardware:

- The IngeniArs SpFi IP core, used inside the SpaceART® link analyser.
- The Cobham Gaisler IP SpFi core, programmed and tested both on a Microsemi RTG4 board and on a Xilinx Virtex-6 board.
- The STAR-Dundee SpFi IP core, used inside the STAR Fire link analyser.

A brief description of the three products is available in Chap. 6.

The tests performed allow to present results in terms of SpFi IP core interoperability. Since it also involved the use of three different link analysers, it provides indications about different features implemented in the Graphical User Interfaces (GUIs) of these devices. The results of the test plan presented in Table 4.1 are then shown in a compact view in Table 4.2. It was not possible to carry out the ML test section as the available CoDecs with link analysers did not support the Multi-Lane layer at that time. Please note that Cobham Gaisler IP core provided by ESA has been referred to as *E*, IngeniArs as *I* and STAR-Dundee as *S*. Consequently, for example, an ESA loopback link is referred to as E–E and an IngeniArs to Cobham Gaisler link as I–E.

As clearly indicated in Table 4.2, no interoperability issues arose: each single SpFi IP core can communicate with each other, exploiting all the main protocol features. However, a series of relevant indicators, which may be useful for both system developer and system users, have been collected:

- Except for the ESA link analyser, neither the STAR Fire nor the SpaceART® currently provides a Lane Reset command on their GUI. (1)
- No error injection mechanism on ESA link analyser, impossible to test. (2)
- IngeniArs and STAR-Dundee implementations consider EOPs and Fills symbols in the bandwidth displayed on the link analyser interface, while the ESA

Table 4.2 Interoperability test results

Test ID	E–E	I–I	S–S	E–I	E–S	I–S
LL-1	**V**	**V**	**V**	**V**	**V**	**V**
LL-2	**V**	**V**(1)	**V**(1)	**V**(1)	**V**(1)	**V**(1)
LL-3	**V**	**V**	**V**	**V**	**V**	**V**
DL-1	**V**	**V**	**V**	**V**	**V**	**V**
DL-2	**NA**(2)	**V**	**V**	**V**	**V**	**V**
DL-3	**V**	**V**	**V**	**V**	**V**	**V**
DL-4	**V**	**V**	**V**	**V**	**V**	**V**
DL-5	**V**	**V**	**V**	**V**	**V**	**V**
QoS-1	**V**	**V**	**V**	**V**	**V**	**V**
QoS-2	**V**	**V**	**V**	**V**	**V**	**V**
QoS-3	**V**	**V**	**V**	**V**	**V**	**V**
QoS-4	**V**	**V**	**V**	**V**	**V**	**V**
QoS-5	**V**	**NA**(3)	**V**	**NA**(3)	**V**	NA(3)
QoS-6	**V**	**V**	**V**	**V**	**V**	**V**
QoS-7	**V**	**V**	**V**	**V**	**V**	**V**

application does not. This results in a mismatch of the displayed bandwidth on the link, inversely proportional to the packet size.

- The IngeniArs link analyser broadcast interface allows sending only a single broadcast message, not to generate them continuously at a configurable bandwidth. (3)
- The timeslot setting has been handled differently in the different IPs: ESA and STAR-Dundee interfaces allow the user to manually select the timeslot of the system, while the IngeniArs automatically ticks the timeslot on a regular time scheme. Since the standard does not specify any particular behaviour, all these configurations are considered perfectly compliant and demonstrated to be interoperable.

In the previous version of the standard, the order on which CRC and scrambling shall be computed was not specified, and thus different developers interpreted it in different ways. In the latest version of the standard (ECSS-E-ST-50-11C), this issue has been handled, establishing an order between the two operations. Consequently, some developers (e.g. ESA and IngeniArs) had to invert the two operations within their IP. This operation has been carried out just before this test campaing, exposing systems to potential bugs insertion. During the test, they have been identified and corrected, obtaining full compatibility and interoperability also on scrambled links.

Chapter 5
Set-Up and Characterisation of a SpaceFibre Network

5.1 SHINe: Simulator for SpaceFIbre and SpaceWire Satellite OBDH Network

SpaceFibre is a very appealing network, but it is also complex to set-up to achieve the desired end-to-end requirements. To help this process, a simulator based on the open-source OMNeT++ [13], named SHINe [10], has been developed. It supports the simulation of the SpaceFibre and SpaceWire protocols to help both the final steps of the standardisation process and the system engineers in the set-up and test of new networks. SHINe allows to precisely simulate common network metrics, such as latency and bandwidth usage, and it can be connected to real hardware in a Hardware-in-the-Loop configuration. With SHINe, it is possible to easily deploy a network via drag&drop from a palette and simulate it, collecting and analysing the results. Being based on OMNeT++, SHINe is completely written in C++, it is easily extensible, allowing the user to develop custom nodes to use together with the existing ones. SHINe implements not only the basic SpaceFibre and SpaceWire protocols, but it also offers a routing switch node, both RMAP target and RMAP initiator nodes, and an advanced Hardware-In-the-Loop (HIL) mechanism to connect physical devices to the simulator.

5.1.1 Software Architecture

The driving idea during the development of SHINe was to create a tool providing the building blocks to easily set up a SpaceFibre or SpaceWire network infrastructure, while the definition of the applications connected to this network is left up to the user. The main two core building blocks provided by SHINe are the SpaceFibre endpoint and the SpaceWire endpoint. They offer a C++ interface compliant with

P. Nannipieri et al., *Next-Generation High-Speed Satellite Interconnect*,
https://doi.org/10.1007/978-3-030-77044-0_5

the service interface described in the two standards, completely hiding the details of the protocol to the application.

As shown in Fig. 5.1, the user-defined application must extend a specific C++ interface to be connected to the endpoints. In such a way, the endpoint can call *callback* functions (a function passed into another function as an argument, which is then invoked inside the outer function to complete some kind of routine or action) of the application without knowing their actual implementation. For example, the endpoint may notify through a callback that new data is ready to be read or that the transmission buffer has a free slot to send new data.

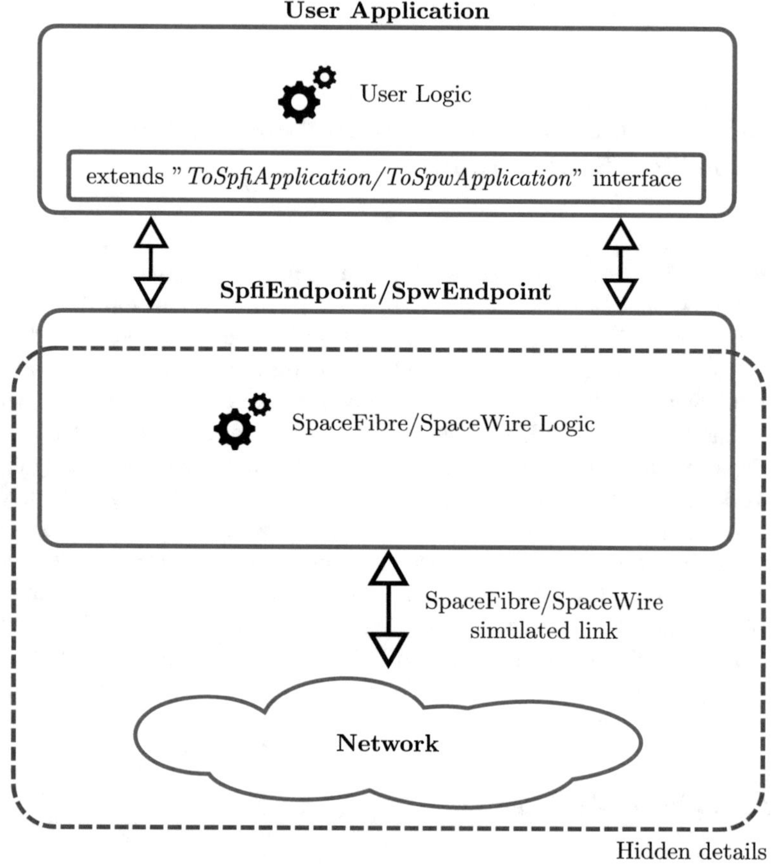

Fig. 5.1 Interaction between the user application and a SpaceFibre or SpaceWire endpoint

5.1.2 SpaceFibre Endpoint

The SpaceFibre endpoint module is the core building block in SHINe to instantiate a fully functional SpaceFibre port. As shown in Fig. 5.2, it provides two input/output gates for the connection with the upper-layer application and one input/output gate representing the physical SpaceFibre connector. Note that there are two gates towards the upper layer, one for the NChar transmission (*appPacket*) and one for the BC messages transmission (*appBroadcast*), allowing using two different applications in case the user prefers to model the two interfaces separately. With NChar transmission we refer to regular data traffic flowing in and out from input and output VCs. In order to be connected to a SpaceFibre endpoint, an application must extend the *ToSpfiApplication* C++ abstract class. This class represents the interface that the endpoint can use to notify about the availability of new NChars or Broadcast messages to read. A SpaceFibre endpoint automatically collects several metrics about network usage, such as (i) packet latency, (ii) packet inter-arrival time, (iii) packet inter-transmission time and (iv) BC message latency.

5.1.3 SpaceWire Endpoint

The SpaceWire endpoint plays a role similar to the SpaceFibre endpoint, being the building block that provides a fully functional implementation of the SpaceWire standard. From the user perspective, the two kinds of endpoints offer a similar interface. A SpaceWire endpoint, whose architecture is shown in Fig. 5.3, provides two input/output gates for the connection with the upper-layer application and one input/output gate representing the physical connector. The connection with the application is split into two gates, one for the NChars transmission and reception and one for the timecodes (BC distributing time synchronisation messages across the network), hence it is possible to use two different applications for the packet stream and the timecodes. Note that interrupts are not currently supported in SHINe.

5.1.4 Routing Switch

The core elements in a network are naturally the routing switches. One of the most critical points in the study and development of the SpaceFibre network layer is its interoperability with legacy SpaceWire networks. A realistic scenario in the transition phase between the two technologies foresees the use of SpaceFibre as a high-speed backbone with some of the nodes connected through SpaceFibre and some others through SpaceWire or other protocols.

An overview of the routing switch architecture is shown in Fig. 5.4. A routing switch comprises a user-defined number of ports of three possible kinds: (i)

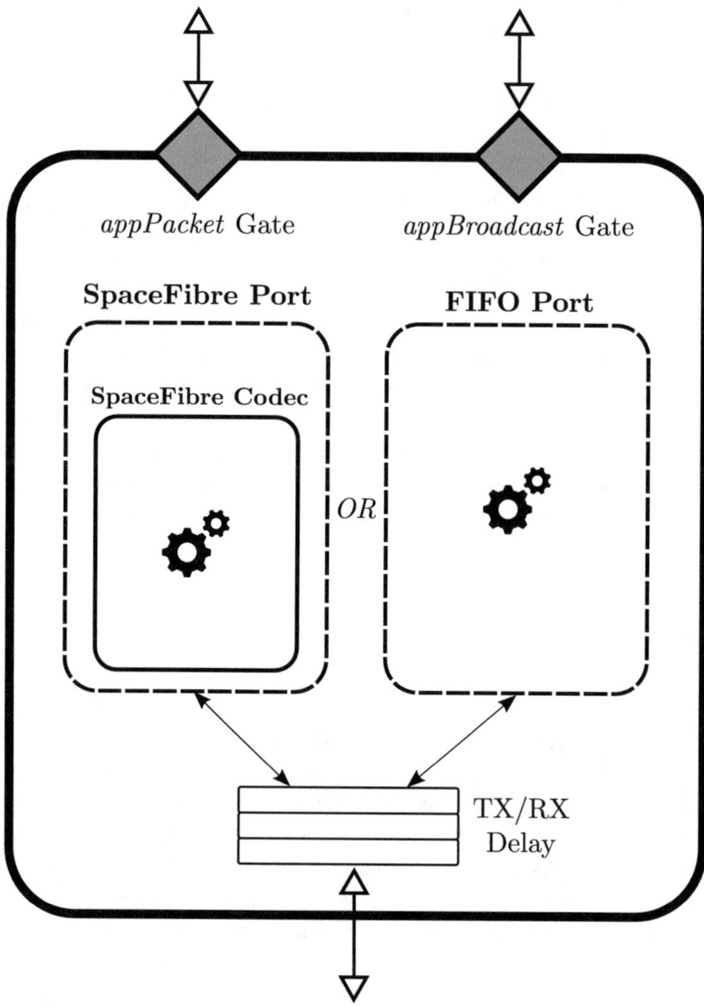

Fig. 5.2 Internal architecture of a SpaceFibre endpoint

SpaceFibre ports; (ii) SpaceWire adapter ports, or (iii) FIFO ports. All of them share the same SpaceFibre interface to the internal switching matrix, but they differ in their implementations. In particular, the SpaceWire adapter port acts as a wrapper around a normal SpaceWire Port, making it look like a SpaceFibre Port with only one virtual channel. Moreover, this wrapper is responsible for the bridging of broadcast messages according to the following rule: whenever a timecode is received from the underlying SpaceWire port, it is wrapped into a BC message of a specific type and then forwarded to the switching matrix to be propagated. Vice versa, whenever the switching matrix tries to propagate a BC message to a SpaceWire adapter port, the message is checked for its type: if it contains a

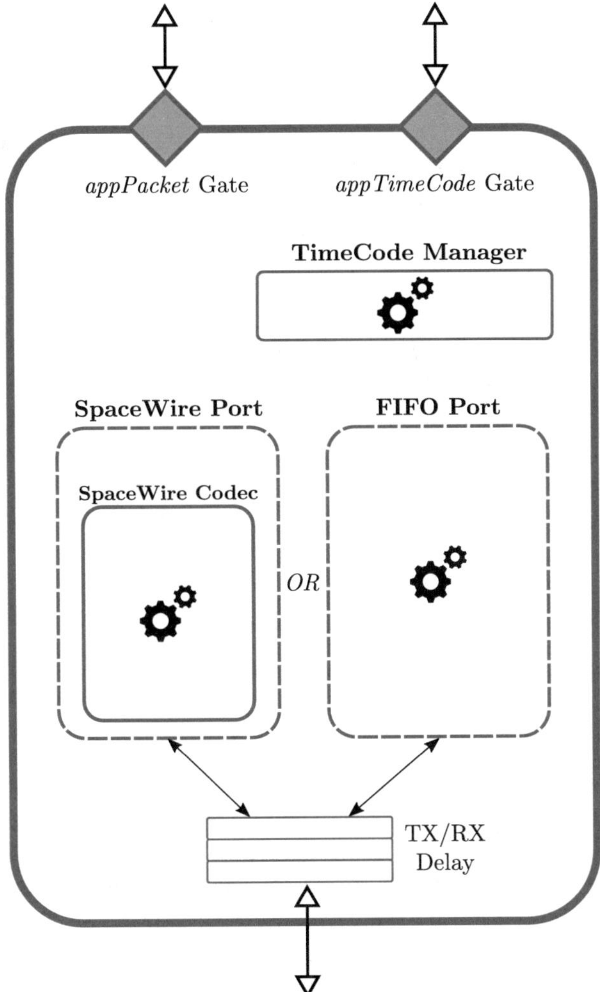

Fig. 5.3 Internal architecture of a SpaceWire endpoint

timecode, the message is unwrapped and the timecode is sent, otherwise it is simply dropped. This simple mechanism makes it possible for a SpaceFibre backbone to transparently carry timecodes. Moreover, the automatic loop prevention foreseen by the SpaceFibre standard guarantees that no multiple copies of the same timecode are propagated. The SHINe routing switch implements all the features required by the SpaceFibre standards. In particular, it supports:

- Path and logical addressing. When using path addressing, the output port of the routing switch is directly written in the packet header. When using

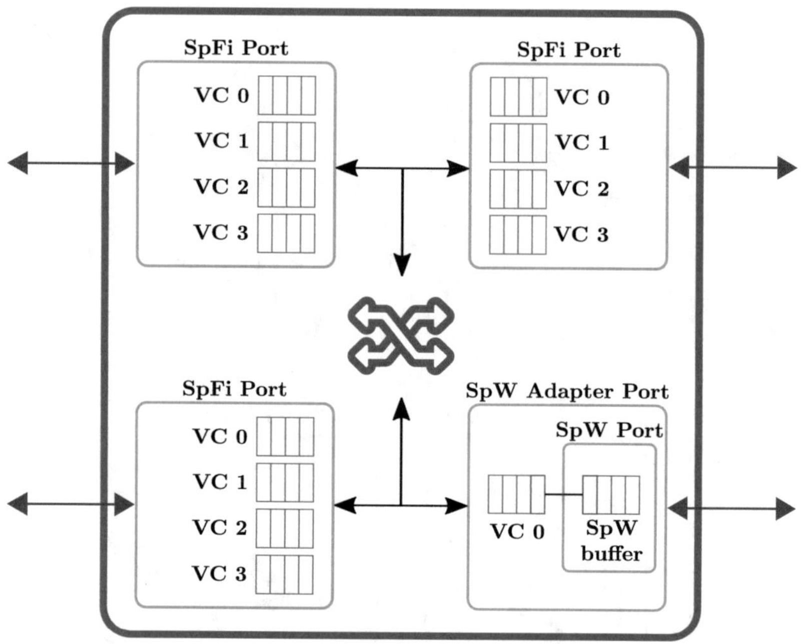

Fig. 5.4 Routing switch with a SpaceWire adapter port

logical addressing, however, a routing table is necessary to associate the logical addresses to the output ports.

- Virtual network mapping. For each virtual channel of each port, the user can specify the virtual network it belongs to. According to the SpaceFibre standard, packets can flow only through virtual channels belonging to the same virtual network.
- Multicast support. In case the user decides to specify a custom routing table, he can define, for a specific logical address, a multicast set of output ports. When multicast is used, an NChar is transferred from the input to all the output ports of the set at the same time, provided that they can all accept it.
- Group adaptive routing support. Similarly to multicast, a custom routing table can include a set of output ports to be used for group adaptive routing. When an input port requires a new wormhole to be open, only the first output port in the set that is found free is chosen, improving network performances.

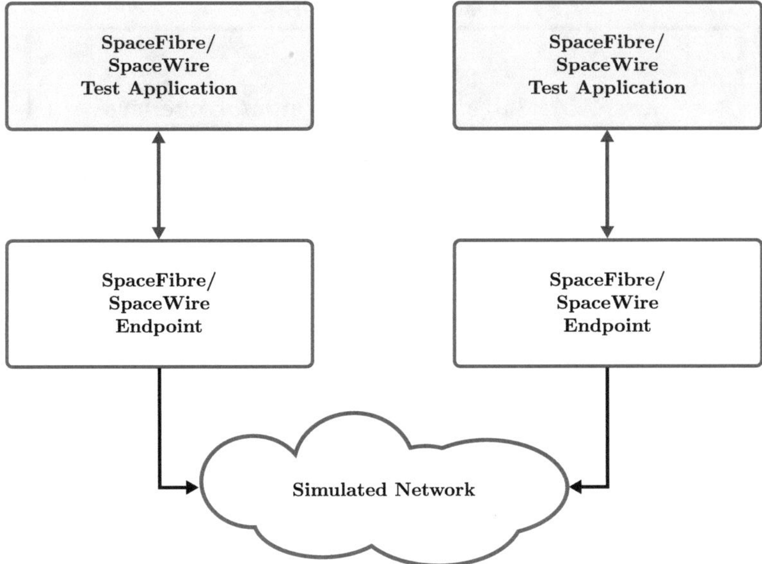

Fig. 5.5 SpaceFibre and SpaceWire test applications

5.1.5 *Test Applications*

SHINe provides out-of-the-box two flexible test applications, one for SpaceFibre and one for SpaceWire, that can be used to study the network performances under different load conditions.

As shown in Fig. 5.5, both the SpaceFibre and the SpaceWire test applications can be directly connected to the correspondent endpoint. They can be configured by the user to simulate a generic device, in particular, it is possible to set: data generation rate and packet size, data consumption rate and destination nodes. The parameters described can be specified individually for each virtual channel. Another possible usage of these example applications is to validate the correctness of the packet: the generated packets contain incremental integers that are checked for integrity on reception.

5.1.6 *RMAP Modules*

The remote memory access protocol [3] is the most common protocol used directly on top of SpaceWire. It allows accessing a remote memory region on the target device in several operation modes. The RMAP standard specifies two types of device: (i) the initiator, which is responsible for issuing the requests and (ii) the target, which must react to the requests and, if necessary, send a reply back to

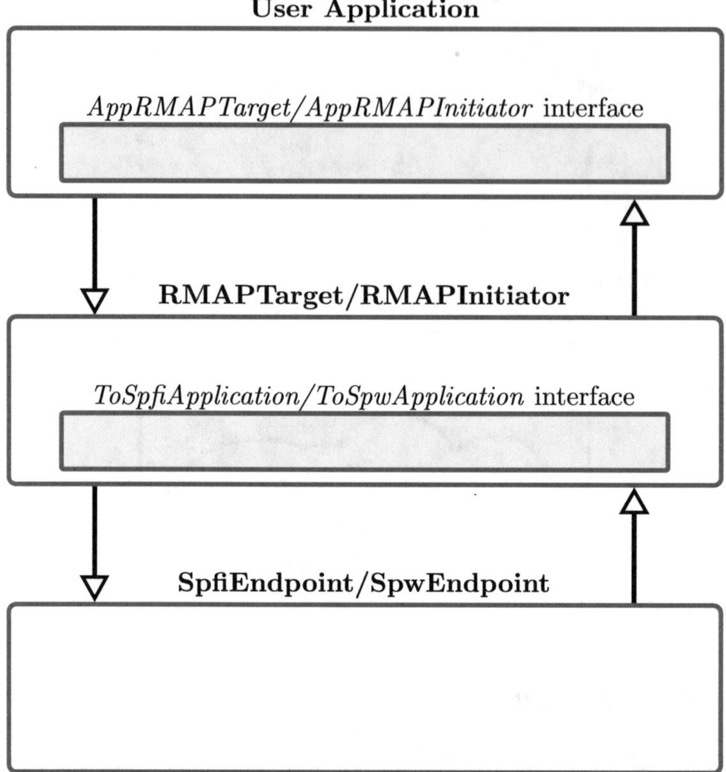

Fig. 5.6 RMAP additional layer between the user application and the endpoint

the initiator. SHINe fully implements the RMAP standard specifications, for both the initiator and the target. They are implemented as application modules for the SpaceFibre and SpaceWire endpoints, actually creating an additional layer between the final application of the user and the network infrastructure (see Fig. 5.6).

5.1.7 Hardware-in-the-Loop

A relevant feature of SHINe is the possibility to connect it in a HIL configuration with IngeniArs SpaceART® [8]. SpaceART is cutting-edge test equipment for the analysis of SpaceFibre and SpaceWire devices, equipped with two SpaceFibre ports, four SpaceWire ports and an Ethernet port for the connection with the host-PC (Fig. 5.7).

Again, the HIL is transparent for both the Unit Under Test (UUT) connected to SpaceART® and for the rest of the simulated nodes in SHINe. Because the simulation speed is much slower than the data rate of a real device, the communication

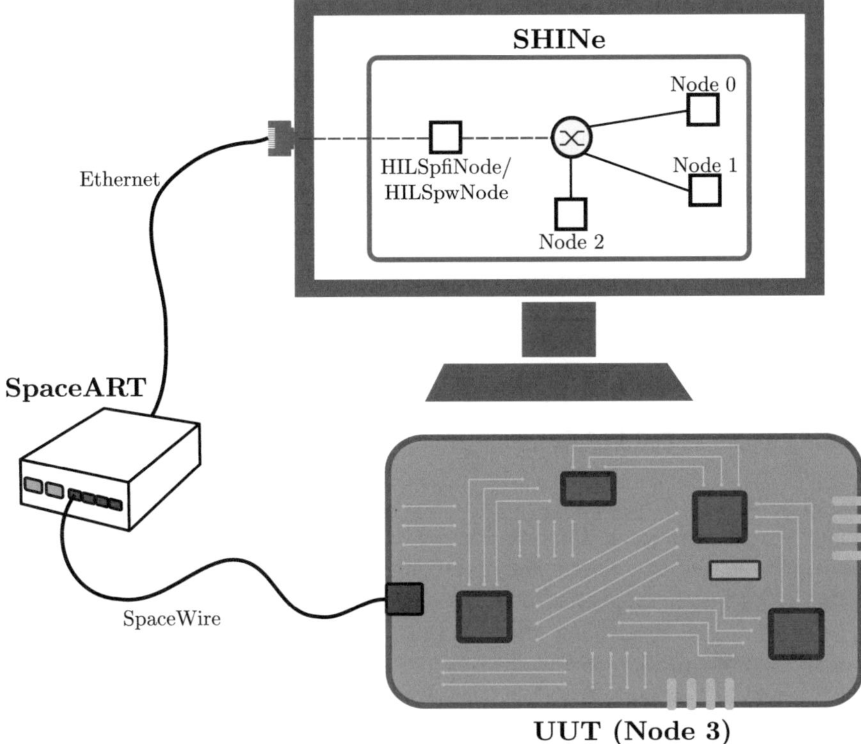

Fig. 5.7 HIL configuration with SHINe connected with SpaceART® and the UUT

will not happen in real-time. This means that the UUT will see a far-end device that is "slow" in consuming and producing data. However, thanks to the flow control mechanisms implemented in both SpaceFibre and SpaceWire, no data will be lost. Note that this is true as long as the UUT does not rely on any time-dependent mechanism such as timeouts, in which case the different time speed might, and probably will, affect the correct behaviour.

5.2 Network Architecture and Configuration

5.2.1 Examples Network Topologies

Before presenting some possible SpaceFibre network topologies, it is important to give a few definitions that will be then used in the entire chapter. We will refer to every single SpaceFibre port or switch as *Node*. A network is composed of a determined number of SpaceFibre ports and SpaceFibre switches. Each port has a

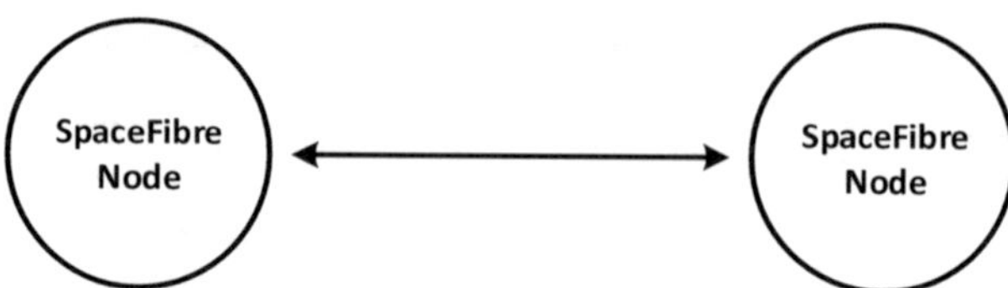

Fig. 5.8 SpaceFibre point-to-point link

single SpaceFibre interface (it may be single-lane or multi-lane), while each switch has two or more SpaceFibre ports and the interconnect logic to perform the routing operations.

5.2.1.1 Point-to-Point Link

The very basic communication architecture achievable with the SpaceFibre technology is a point-to-point link (see Fig. 5.8). It is the technology most likely to be adopted in the nearest future: it does not require any routing switch, but only SpaceFibre CoDec interfaces. At the time of writing, a technology demonstrator mission employing several SpaceFibre points to point is under development: the ISS-TAS Orbital Experiment was scheduled for launch in Q1 2020, with several SpaceFibre implementations working together. The scientific community is currently waiting for reports about that mission, as it will be the first mission with the SpaceFibre interface on-board.

5.2.1.2 Network: Star Topology

The star topology is the less complex networked SpaceFibre topology. It is made up of a SpaceFibre routing switch connected to several SpaceFibre ports. In Fig. 5.9 an example with 5 SpaceFibre ports is shown. Such a topology potentially enables the communication between all the nodes in the network, passing through the router. The simple star topology has few drawbacks: vulnerability first of all. In fact, in the case of a routing switch failure, the entire network would go down. Even though it will not be probably used in future missions for its reliability limitations, thanks to its simplicity, it will be useful to perform tests on the network itself, as it is shown in Sect. 5.2. The achieved results can be then easily scaled to more complex topologies.

5.2.1.3 Network: Double Star Topology

A double star topology begins to fits more closely with the real needs of a SpaceFibre OBDH network. Such topology may take advantage efficiently of the SpaceFibre protocol features: we can think of a scenario where several instruments

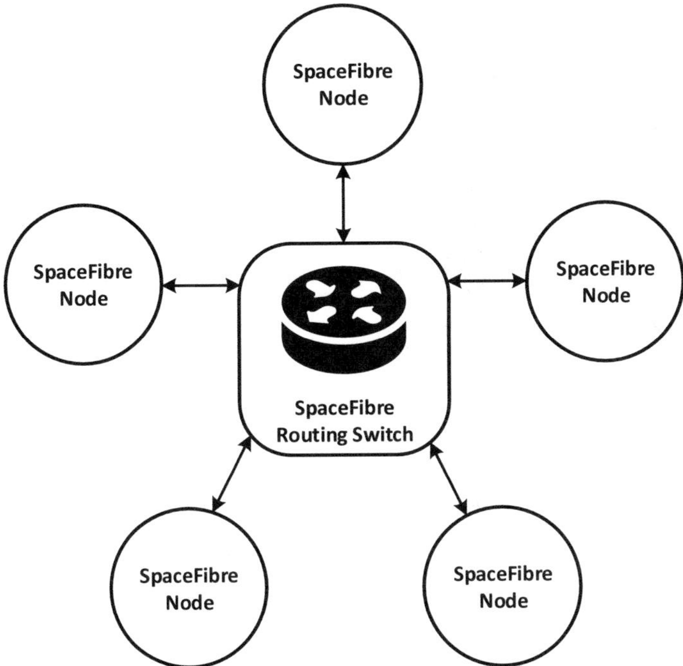

Fig. 5.9 SpaceFibre network: Star topology

within the payload are connected to a first routing switch and need to be connected to several nodes in a separate section of the spacecraft where platform operations are made (e.g. data processing, storage, control, telemetry). In such a situation the router, possibly equipped with Multi-Lane (to add redundancy on critical links) SpaceFibre CoDecs, can act as data concentrators, encapsulating several virtually separated data flows on the same physical link travelling through a separate section of the spacecraft. It is for such a reason that SpaceFibre has been recently added in the SpaceVPX VITA 78.1 standard [6], where the SpaceFibre protocol is seen as a useful backplane interconnect (Fig. 5.10).

The presented topologies just represent the very basic ones: it is possible to adapt the network topology to specific mission requirements by combining and modifying them. For example, by adding more hops to the network, it is possible to implement QoS and FDIR not only at the point-to-point level but also at the network level. To realise that, an upper protocol running on top of the network layer is necessary. Such a layer, known as the "Transaction layer" is currently under ESA development and will be the next main research topic within the SpFi community.

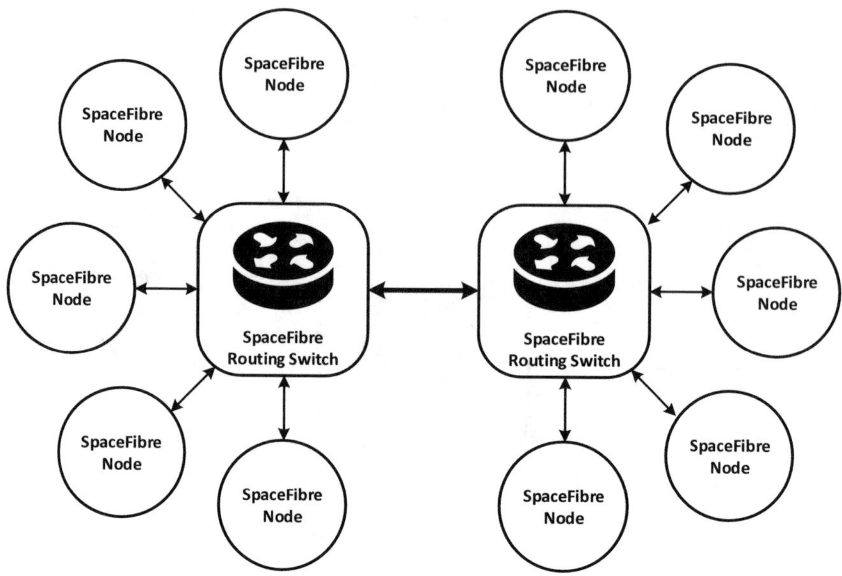

Fig. 5.10 SpaceFibre network: Double star topology

5.2.2 Representative Network Design

In the following, we present a representative SpaceFibre network, specifically designed to be considered a realistic representation of an on-board data-handling network. Link analysers and CoDec configurators are employed to emulate all the different building blocks of payload data-handling systems. This allowed us to perform a series of functional and performance tests which will be presented in Sects. 5.3 and 5.4. This analysis can be useful for potential technology adopter to understand the strengths, weaknesses and peculiarities of the SpaceFibre network technology. The controlling idea is to study, assess and evaluate performances and properties of a generic network, to produce results easily understandable, which may be applied then to other more specific network topologies. The SpaceFibre network chosen to accomplish the task is the star topologies, where each device in the network is connected to any others through a single router, as already shown in Fig. 5.9. The obtained results characterise both point-to-point and routing switches performances, therefore they can be applied to more complex network topologies to provide an estimation of their timing performances. It is remarkable to say that the results we are seeking here are different concerning the one that can be obtained with a network simulator like SHINe: the network simulator can assess a very complex network from a high-level point of view: network QoS, advanced statistics, and other metrics. However, being a simulator, the results SHINe provides are just an estimation. To have a solid and reliable measure of the timing performances, it is necessary to perform real hardware tests, which will be bounded to the specific

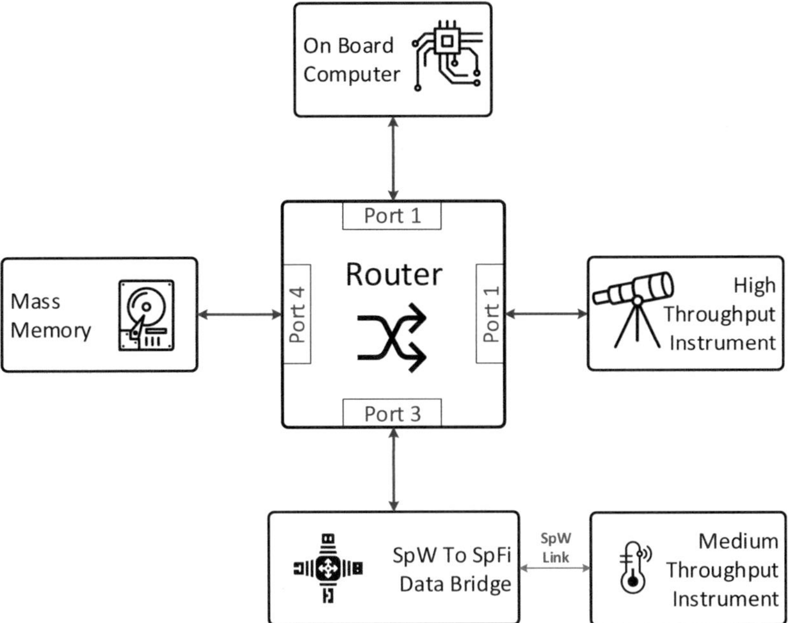

Fig. 5.11 Payload data-handling representative network

hardware employed. Latency, jitter and other figures of merit are closely related to the specific hardware used and the only way to have a real estimation of them is to directly measure them. Figure 5.11 shows the block diagram of the emulated network. It is composed of a routing switch with four SpaceFibre ports, connected to the following component:

- Port 1 is connected to the on-board computer with a SpaceFibre communication link;
- Port 2 is connected to an instrument producing a high amount of data (>1 Gbps) through a SpaceFibre communication link;
- Port 3 is connected to the instrument producing a medium amount of data (<200 Mbps) through a SpaceWire communication link bridged on a SpaceFibre link, which mixes up SpaceFibre and SpaceWire traffic;
- Port 4 is connected to the mass memory with a SpaceFibre communication link;

The interconnections done by the routing switch exploit the VN mechanism: each tuple <Port-VC> is part of a VN of 2 or more ends, which is a subset of the real network and is real-time configurable by the user through RMAP commands. Details and references on this mechanism are already given in Sects. 2.6 and 3.3. The designed network, composed of several VNs, is depicted in Fig. 5.12 and detailed in Table 5.1.

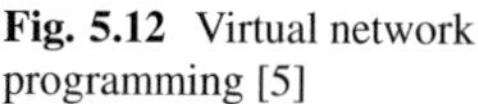

Fig. 5.12 Virtual network programming [5]

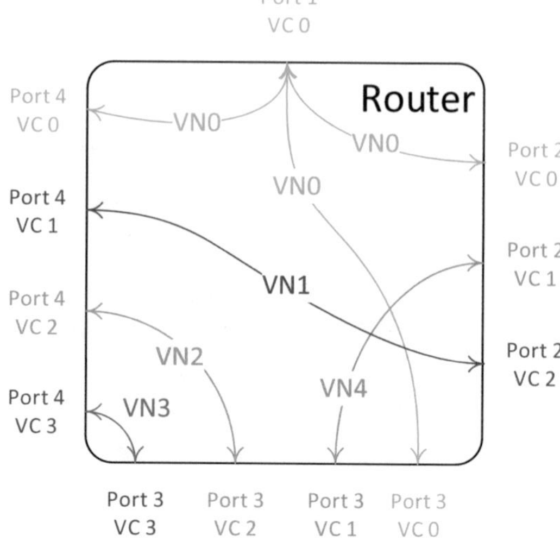

Table 5.1 VN description

Virtual network	Requested bandwidth	Priority level
VN0: OBC control traffic	<1%	1
VN1: High-bandwidth instrument traffic	65%	2
VN2: Medium bandwidth instrument SpFi traffic	15%	3
VN3: SpW traffic bridged on SpFi	5%	4
VN4: Inter-instrument communication link	10%	5

The network, which is the set of VNs, has been designed to withstand the specific requirements of each device:

- VN0 is the VN with the highest priority. It interconnects all the devices with very low bandwidth reserved. It is meant to be used for control commands sent from the OBC over the network. Such commands do not require significant bandwidth but shall be delivered with maximum priority and minimum delay. The use of VN0 is also prescribed by the standard, where it is specifically required that all the ports are connected to the VN0 with only their VC0;
- VN1, VN2 and VN3 are the real data-handling VN. They support several classes of data traffic with different bandwidth and priority, all generated by the instruments and going to the mass memory;
- VN4 is a VN that allows an inter-instrument communication link. Such links can be used for synchronisation, calibration or any other non-critical control purposes. We represented them with the lowest priority level and low bandwidth requirements;

Please note that in our network we refer to 1 as the highest priority level, 5 as the lowest priority level: the higher the number is, the lower the priority level is.

The exact configuration parameters for all the VNs are detailed in Table 5.1. The requested bandwidth is derived from the maximum requested bandwidth for every single SpFi link, ensuring that none of these exceeds 95% (leaving bandwidth for protocol control word exchange) of the full bandwidth:

- SpFi Port 1: $\leq$ 1%.
- SpFi Port 2: 75%.
- SpFi Port 3: 25%.
- SpFi Port 4: 85%.

Each device of the network is emulated employing both EGSE available on the market and IP cells mapped onto the FPGA development board.

5.2.2.1 Hardware Resources

The building blocks of the network described in Fig. 5.11 have been chosen within three different providers of SpaceFibre technologies: IngeniArs, STAR-Dundee and the Cobham Gaisler, provided by ESA IP pool for R&D purposes.

Router: IngeniArs Routing Switch IP

The routing switch [9] is owned and provided by IngeniArs and mapped on a commercial Xilinx Ultrascale+ ZCU102 board [15], equipped with appropriate high-speed serial connectors thanks to the ALDEC SATA FMC module [1]. The board is shown in Fig. 5.13.

The SpaceFibre ports are implemented with the IngeniArs SpaceFibre CoDec IP [2]. A key feature of the routing switch is that it can be instantiated with a generic number of ports, each one with a generic number of VCs. For a detailed description of the routing switch, refer to Sect. 3.3 and [9].

Mass Memory: IngeniArs SpaceART

IngeniArs SpaceArt, already described in Sect. 3.4, is used for emulating the behaviour of mass memory, by saving the received data on external memory support.

High-Bandwidth Instrument and SpW over SpFi Bridge: ESA Owned SpaceFibre CoDec IP

The high-bandwidth instruments have been emulated employing the Cobham Gaisler SpaceFibre IP core available from the ESA IP pool, implemented on a Microsemi RTG4 board. The same IP core has also been used on a Virtex-6 board [16] equipped with the STAR-DundeeSpW/SpFi FMC board [14]. In this

Fig. 5.13 SpaceFibre router on a Xilinx ZCU102 board with ALDEC FMC expansion board

configuration, the incoming SpaceWire traffic is directed into one of the VC of the SpFI IP core and from there sent over the network through the SpaceFibre port. The IP core is described with more details in Sect. 6.2.

On-Board Computer: STAR-Dundee StarFire

The OBC has been emulated employing the MK2 StarFire from STAR-Dundee, described with more details in Sects. 6.2 and 6.4.

Medium Bandwidth SpaceWire Instrument: STAR-Dundee SpaceWire Brick

The SpaceWire traffic has been generated and sent over the network by the SpaceWire brick. It is a two SpaceWire and one USB port device, interfaced with the host-PC, able to generate SpaceWire traffic. It does not have advanced features such as link analysis or error injection, thus it is used in combination with a SpaceWire link analyser [11].

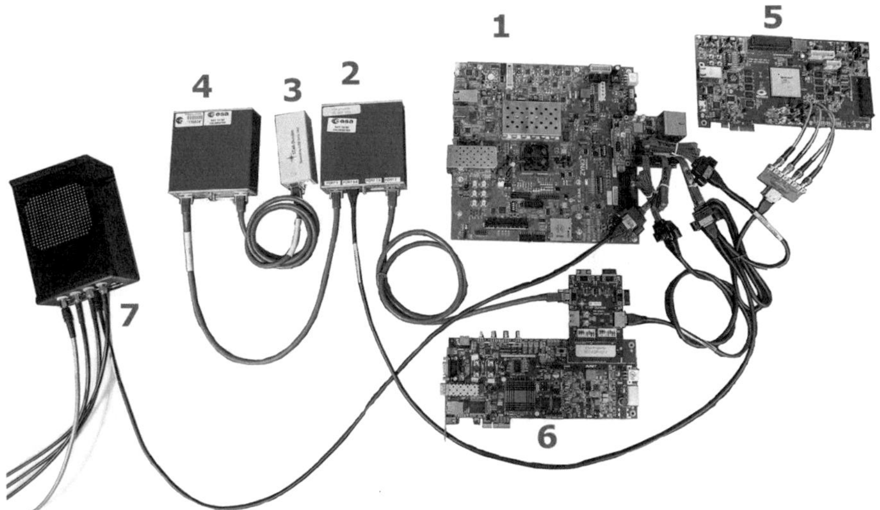

Fig. 5.14 SpaceFibre network prototype under test [5]

Assembled Network

Up to the author's knowledge, this work also presented in [5] is the first literature contribution about a SpaceFibre network realised combing different devices from different manufacturers. Besides the assembly of the network itself, a series of tests have been carried out, to assess the functionalities and performances of the system. In Fig. 5.14 an overview of the assembled network is given, with:

- Device 1: the routing switch, implemented on the Xilinx ZCU102 Board.
- Device 2: the OBC emulator, realised using SpFi port 1 of the StarFire from Star-Dundee.
- Device 3: the medium bandwidth SpW instrument, which generates SpW traffic going through device 4 (a SpW link analyser) and through device 2 which adds error injection possibilities on the data flow; the SpW data flow finally gets to device 6, where it is multiplexed on a VC o a SpFi link directed to the routing switch.
- Device 5: the high-bandwidth instrument emulator, with the Cobham Gaisler SpFi IP core provided by ESA and link analyser implemented on a Microsemi RTG4 Board.
- Device 7: the mass memory emulator, realised using port 1 of the SpaceART® from IngeniArs.

5.3 Functional and Performance Test Plan

5.3.1 Functional Tests

The assembly of a SpaceFibre network can be verified by executing a hardware tests suite, to guarantee the system correct behaviour. In the following, we will present a prototype test plan, which we used to assess our build network but can be adapted to any other generic network. Please note that we assumed that interoperability at a point-to-point level is already ensured by an adequate test campaign, as shown in Chap. 4. Indeed, tests focused on packets and broadcast transmission under various conditions. Packets have been generated exploiting the Host interfaces and the link analysers tool provided by the various endpoints. Incremental data is generated and encapsulated in packets so that the link analysers tool can easily check for data packets integrity at their receiving end. Also, broadcast messages are generated and sent by each node to the rest of the network, which shall verify the correct reception. The hardware tests focused on priority mechanism and bandwidth reservation at the network level. We did not include in our test any timeslot scheduling mechanism at the network level since different devices do not share the same approach in timeslot configurability. The test plan shall include intensive packets and broadcast communication with error injection. In particular, the system demonstrates to be able to communicate with BER up to 10^{-5} (according to what the standard says), obviously encountering a massive (−70/80%) link efficiency reduction. However, when operating with typical BER value (10^{-12}) [17] we did not record significant changes in the link efficient bandwidth. In Table 5.2, we present a summary of a functional reference test plan. Please note that it extends the test plan already presented in Chap. 4 and [5], without the Multi-Lane support since at the time of writing routing switches do not support Multi-Lane communication yet.

5.3.2 Performance Tests

The first step in the performance analysis of a SpaceFibre network shall be the detailed description of the figures of merit that shall be measured. In the following, the definition proposed in [5].

- **Broadcast Latency** (Δ_{BC}): time interval between the writing of a BC message in the near-end TX BC interface and the reading of the same BC in the RX BC message interface of the far-end.
- **Broadcast Jitter** (δ_{BC}): time interval uncertainty on the Δ_{BC} measurements.
- **Packet Latency** (Δ_{PKT}): time interval between the writing of the first word of a packet in the near-end OUT VCB and the reading of the packet EOP in the IN VCB of the far-end.
- **Packet Jitter** (δ_{PKT}): time interval uncertainty on the Δ_{PKT} measurements.

Table 5.2 Functional analysis test plan [ref]

Test ID	Test description
DL-1	Verify that full bandwidth communication occurs on the entire network as specified in Table 5.1 without any loss of data/nacks
DL-2	Verify that full bandwidth communication occurs on the entire network as specified in Table 5.1 without loss of data when injecting errors at Lane layer level
DL-3	Communication with scrambling activated/deactivated independently on various ports
DL-4	Broadcast communication test across the entire network from each single port
DL-5	Verify that no data loss occurs if cables are manually disconnected during normal communication
NL-1	Full communication on the entire network as specified in Table 5.1 with small packet sizes [1–50]
NL-2	Full communication on the entire network as specified in Table 5.1 with medium packet sizes [51–1000]
NL-3	Full communication on the entire network as specified in Table 5.1 with big packet sizes [9000]
NL-4	Verify broadcast transmission does not exceed maximum allowed bandwidth value
NL-5	Verify that bandwidth reservation mechanism is correctly implemented and is compliant to the values described in Table 5.1
NL-6	Verify priority setting is correctly implemented and respected all over the network

- **Word Latency** (Δ_W): time interval between the writing of the first word of a packet in the near-end OUT VCB and the reading of it in the IN VCB of the far-end.

To measure and characterise the above-mentioned parameters, we exploit the two-node star topology network shown in Fig. 5.15. It is made of two SpaceFibre nodes communicating through a single routing switch.

According to the definitions presented above, we need to measure with a negligible error very short time intervals occurring between conditions that are directly observable at high-level. The strategy proposed in [5] prescribes that the nodes of the network shall be customised to output pulses on external trigger connectors whenever a significant condition occurs. By doing that the test operator will be able to measure time intervals between significant events with the help of an oscilloscope. The drawback of this approach is that you need to have access to the RTL code of the SpaceFibre interfaces and to modify the circuits creating the required output pulses. In our performance test campaign, thanks to the support of ESA and IngeniArs, we had the chance to modify their IP and consequently to characterise their performance within a network. We configured the SpaceFibre IPs to produce a single clock cycle output pulse each time that at the upper interface of the Data-Link layer of the SpaceFibre IP:

- A broadcast message was sent or received.
- The first word of a packet was sent or received.
- The (EOP) word was sent or received.

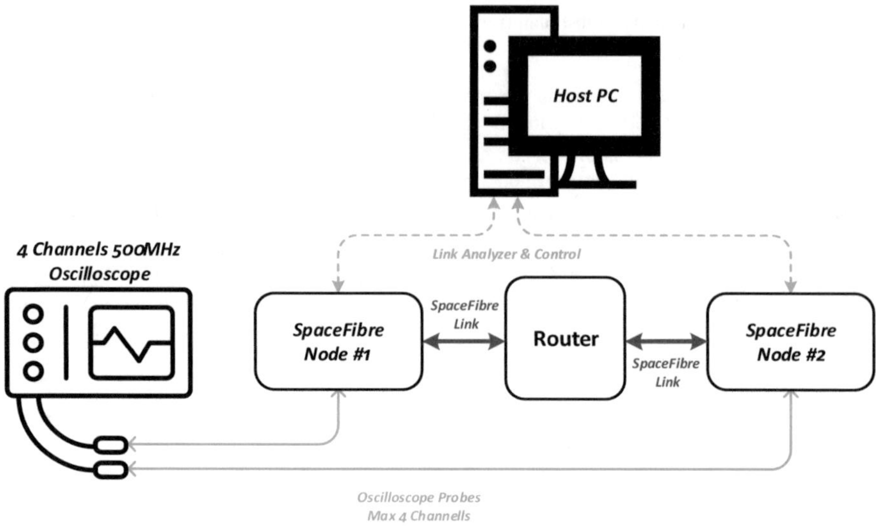

Fig. 5.15 Performance analysis test set-up [5]

To evaluate the network performances in terms of latency and jitter for both data packets and broadcast messages, the network underwent a series of tests.

Packets

Concerning data packets, it is remarkable to say that both latency and jitter over a network are intrinsically dependent on the status of the link. The characterisation work we are proposing aim at deriving a latency and jitter estimation model, varying the packet size and the link status. The controlling idea is to extrapolate a formula that will mix device-dependent and protocol-dependent parameter to build a latency/jitter estimator tool. We focused on the following network and link variables:

- **Packet size:** tests are carried out varying the packet size. They are generated, respectively, with 1, 10, 100 frame size, which is equivalent to 64, 640 6400 word size, considering that a frame is made of 64 SpaceFibre words.
- **Bandwidth reservation:** test is carried out also varying the bandwidth reservation. In particular, we operated with the link fully reserved for only one VC, with two competing VC having 50% reserved bandwidth each and with one VC having 99% bandwidth reserved and a second VC with only 1% bandwidth.

The traffic condition has been chosen to span over all the possible communication scenarios. However, it shall be pointed out that measurements with 1% bandwidth suffer from limited accuracy. Indeed, CoDecs bandwidth configuration accuracy is 0,5%, which means that when we set it to 1%, the actual bandwidth occupation is between 1,5% and 0,5% of the available bandwidth. Since we aim to provide an empirical estimation of latency and jitter over a SpaceFibre network, this is not considered as a blocking point: the obtained results will be significant even with this limited uncertainness.

We do not advise introducing other variables in the performance test campaign. It is true that also other values will affect the measured parameters, the BER above all, but at the same time, they will analyse the results more complex. Because we are aiming at an estimation model, such parameters can be derived separately and then applied to the model as a correction factor. What is important is to have a model that characterises the behaviour of a SpaceFibre network under nominal conditions.

Broadcast Messages
The standard [4] specifies that latency and jitter of broadcast messages shall not be dependent on the bandwidth already occupied on the link, since their priority level is greater than the one of any data packets. Therefore their latency and jitter shall be evaluated without varying the bandwidth allocation parameters. Somehow, the broadcast latency an indirect measure of the minimum link latency, physically bounded to the pipeline and synchronisation register that the broadcast message has to pass through to be delivered to the far-end.

5.4 Performance Test Campaign Results

In this section, we report and analyse the results obtained in our network performance characterisation. Concerning the functional tests presented in Sect. 5.3.1, we employed the routing switch provided by IngeniArs and the CoDecs from Cobham Gaisler (provided by ESA for R&D purposes), IngeniArs and STAR-Dundee. The test plan has been executed on the representative network presented in Sect. 5.2.2. All the tests have been executed by configuring the devices in the network and monitoring the link status and data traffic, thanks to the available link analysers. In particular, thanks to the link observability guaranteed by the link analysers, the tests have been verified by visual inspection of the data stream and link status parameters.

For what concerns the performance test campaign, it has been more detailed and the obtained results needed a deeper analysis. As mentioned in Sect. 5.3.2, it is necessary to have internal access to the RTL of the component of the network and to perform small modification to carry out the prosed test campaign. For this reason, we tested the performance of a network composed of the IngeniArs router, the IngeniArs CoDec and the Cobham Gaisler CoDec, which we had available for modifications. The same test routine can be applied also to other CoDec with the same RTL modification, hopefully with comparable results. It is relevant, for the sake of data soundness and for future comparison, to specify the environmental test conditions. In particular, the virtual channel size of the IPs involved was fixed at 512 data words both for input and output VC buffers. Connections have been realised with a 1-meter eSata crossed cable. Measurements have been taken with a 500MHz oscilloscope (TDS5054B). The data rate of each tested link is fixed at 2.5 Gbps, which means that the clock operating frequency is fixed at 62,5 MHz.

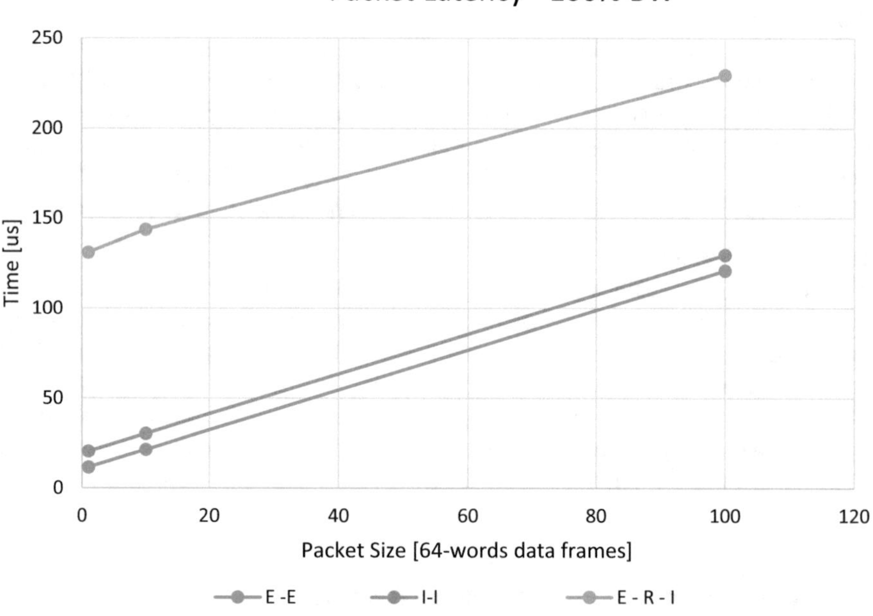

Fig. 5.16 Performance analysis: Packet latency & jitter, 100% reserved [5]

The IngeniArs CoDec is the one available in the SpaceART® link analyser, while the Cobham Gaisler CoDec has been mapped, together with its link analyser, on a Virtex 6 board. These test results have been presented in [5].

Packets

In Fig. 5.16, the packet latency Vs the packet size of a fully reserved link established over the network is shown. In Fig. 5.17, the latency is presented for the 50% reserved link and, in Fig. 5.18, for the 1% reserved link. In the following figures, we refer to E = Cobham Gaisler Owned CoDec IP provided by ESA, I = IngeniArs Owned CoDec IP, R = IngeniArs owned Router IP. The test set-up is described as combining letters, representing each link.

No jitter is reported in the figures, as its value is always lower than 0,5% of the total latency, and can thus be neglected. We can observe that in all the cases latency linearly grows with packet size: of course, the transfer of a bigger packet takes longer. It is also noticeable that for a very small packet the latency approaches the value Δ_W, which has been independently measured. Therefore the following relation can be derived:

$$\Delta_{PKT} = \Delta_W + m * Packet_{size} \tag{5.1}$$

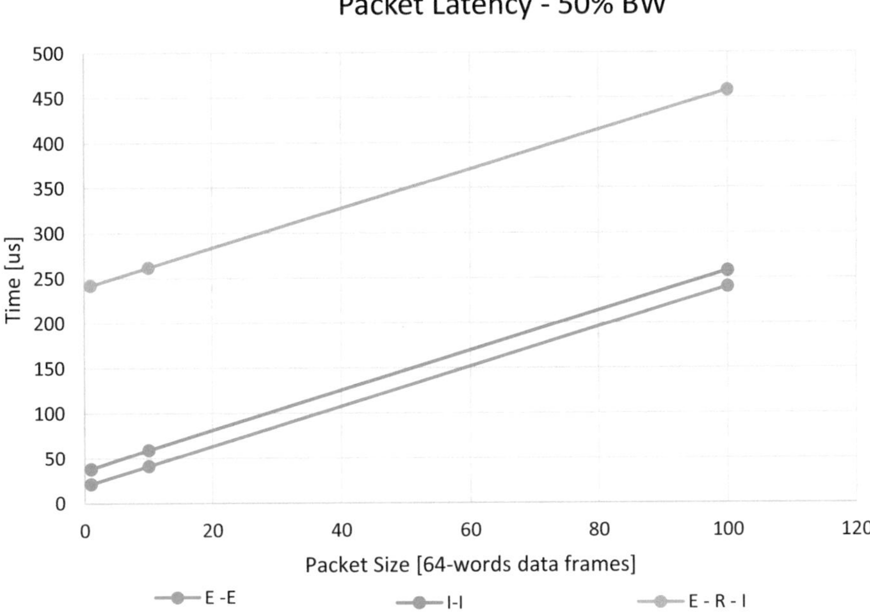

Fig. 5.17 Performance analysis: Packet latency & jitter, 50% reserved [5]

The measured values of $m[\mu s/Pkt_{size}]$ are shown in Table 5.3. It is clearly observable in Figs. 5.16, 5.17 and 5.18 that it is implementation-independent, thus directly related to the SpaceFibre standard mechanism. Moreover, we can also observe that it is inversely proportional to the bandwidth itself.

The latency of a generic SpaceFibre link can be therefore derived with the following relation:

$$\Delta_{PKT} = \Delta_W + Packet_{size}/Bandwith[\%] \tag{5.2}$$

where only Δ_W is an implementation-dependent parameter. These empirical results can be interpreted straight forward: the transmission time for a packet, besides a fixed value for the link latency, is inversely proportional to the bandwidth expressed in percentage allocated for the communication and inversely proportional to the packet size. The largest percentage of the allocated bandwidth will be, the shortest packet transfer time. The bigger the packet will be, the longer it will take to transfer it over the link. These results are very promising in terms of performance if we compare them with state-of-the-art solutions. In [7], an extensive analysis on the timing performance of a SpaceWire network is carried out, with the outcome that for single-hop networked communication, time per single bit transferred is in the order of $10[\frac{ns}{bit}]$, with this value not strongly affected by the packet size. In the single-hop SpaceFibre network analysed, we obtained much more varying results:

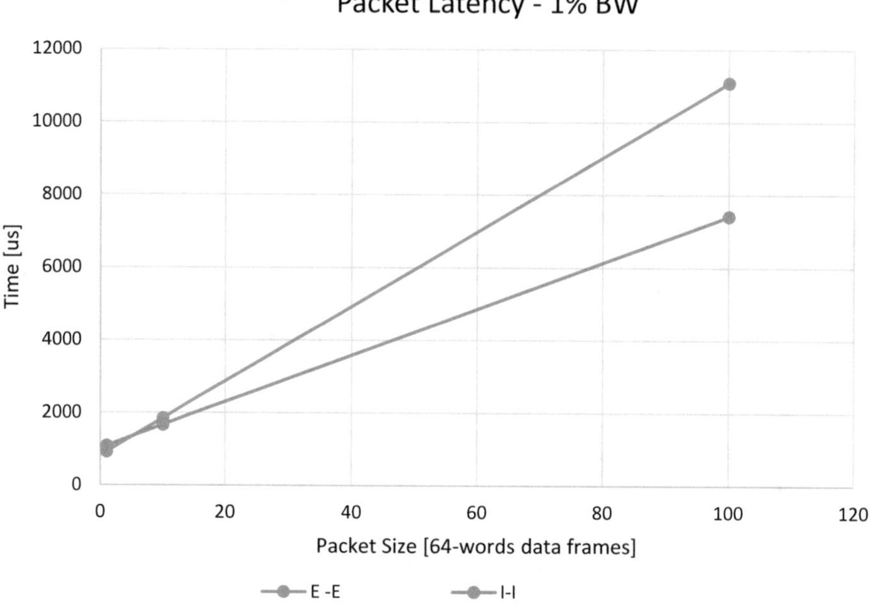

Fig. 5.18 Performance analysis: Packet latency & jitter, 1% reserved [5]

Table 5.3 Packet latency linear coefficient [5]

Reserved bandwidth	m [$\mu s/Pkt_{size}$]	1/Reserved bandwidth
100	1,1	1
50	2,2	2
1	102	100

for small packets, due to the lower link efficiency together with the delay of the router to correctly establish the communication channel, we obtain value in the order of $50[\frac{ns}{bit}]$. However, this value is strongly reduced for much bigger packets, going up to $1[\frac{ns}{bit}]$. This is due to the much large available bandwidth (200 Mbps vs. 2.5 Gbps). Even bigger discrepancies arise with other state-of-the-art data-handling protocols, like TTEthernet; in the work presented in [12], the time per single bit transferred value is in the order of $80[\frac{ns}{bit}]$. Moreover, consider that these values refer to a 2.5 Gbps SpFi link, which is less than half of the maximum achievable bandwidth (6.25 Gbps).

Broadcast Messages

In Fig. 5.19 latencies and jitter measurements for broadcast messages tests are shown.

We observe that the latency of a broadcast message is about 0,6μs for the point-to-point link and 1,4μs for a BC transmitted through a router. In particular, given the

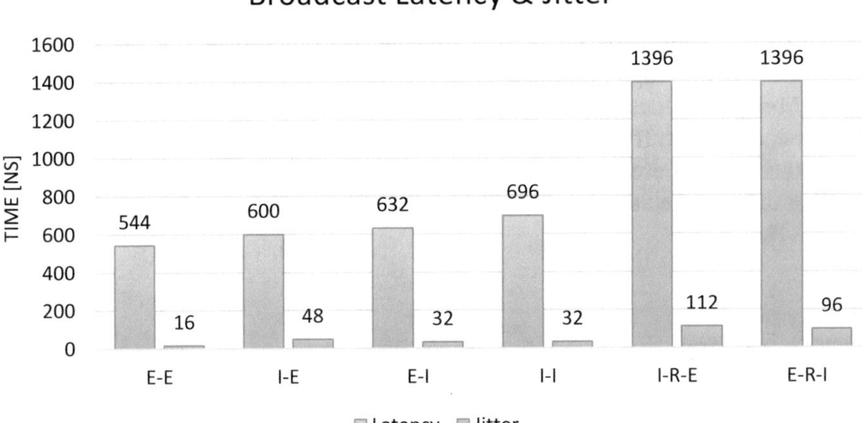

Fig. 5.19 Performance analysis: Broadcast latency & jitter. E = Cobham Gaisler Owned CoDec IP provided by ESA, I = IngeniArs Owned CoDec IP, R = IngeniArs Owned Router IP. Test set-up is described combining letter, representing each link

latency of the IngeniArs to IngeniArs CoDec point-to-point link $\Delta_{I-I} = 696$ ns, the latency of the IngeniArs to Cobham Gaisler CoDec point-to-point link $\Delta_{E-I} = 632$ ns, and the latency of a Cobham Gaisler to IngeniArs (through a routing switch) link Δ_{E-I-R}, we can derive the latency introduced by the router Δ_R. Indeed, it is composed of two point-to-point links plus the routing logic:

$$\Delta_{E-I-R} = \Delta_{E-I} + \Delta_R + \Delta_{I-I} = 1,396\ \mu\text{s} \tag{5.3}$$

$$\Delta_R = \Delta_{E-I-R} - \Delta_{E-I} - \Delta_{I-I} = 68\ \text{ns} \tag{5.4}$$

Considering that the clock period is 16 ns, the latency introduced by the router is only a few clock cycles. For what concern the jitter, we observe that it is always a multiple of the clock period: depending on how a SpaceFibre IP is built and configured, it may have several clock cross-domains crossing on the data path. Some of them are unavoidable to be compliant with the standard, others are user-configurable. Anyway, each of them contributes with a single clock cycle jitter. In a standard SpaceFibre interface the potential source of jitter is the following:

- RX elastic buffer.
- Skip word insertion.
- Optional TX Buffer, usually not bypassed in Xilinx based implementation.
- Output and input virtual channel buffers.
- Router BC forwarding.

Considering that each of these potentially contributes to one clock cycle of jitter to broadcast messages transmission, system designers can configure appropriately the ports to obtain the desired values of jitter. For instance, using the same clock signal for the SpaceFibre interface and upper-layer protocols results in avoiding clock cross-domain in the VC buffers. The values of jitter measured are compliant with the list: the slight difference between the links lays in the fact that the SpaceFibre ports clock distribution scheme has been configured differently by the designers of the two IPs. For more details on the SpaceFibre protocol and to deeply understand what these sources of jitter (buffers, SKIP words, etc.) represent and whether they can be taken away or not, please refer to [4].

References

1. Aldec. Fmc-net networking daughter card. https://www.aldec.com/en/products/emulation/daughter_cards/fmc_daughter/fmc_net
2. Dinelli, G., Marino, A., Dello Sterpaio, L., Leoni, A., Fanucci, L., Nannipieri, P., & Davalle, D. (2020). A serial high-speed satellite communication codec: design and implementation of a spacefibre interface. *Acta Astronautica, 169*, 206–215.
3. European Cooperation for Space Standardisation. (2010). *RMAP Standard ECSS-E-ST-50-52C*. European Cooperation for Space Standardisation.
4. European Cooperation for Space Standardisation. (2019). *SpaceFibre – Very high-speed serial link, ECSS-E-ST-50-11C*. European Cooperation for Space Standardisation.
5. Fanucci, L., Nannipieri, P., & Siegle, F. (2020). A representative SpaceFibre network evaluation: Features, performances and future trends. *Acta Astronautica, 176*, 313–323.
6. Florit, A. F., Villafranca, A. G., McClements, C., Parkes, S., & Srivastava, A. (2017). A prototype SpaceVPX lite (vita 78.1) system using SpaceFibre for data and control planes. In: *2017 IEEE Aerospace Conference*, pp. 1–9. IEEE.
7. Guo, L., Cao, S., & Chen, X. (2010). Research on the delay jitter performance of SpaceWire network for space applications, vol. 5. In: *2010 International Conference On Computer Design and Applications*, Qinhuangdao, China, pp. V5235–V5241.
8. Leoni, A., Dello Sterpaio, L., Davalle, D., & Fanucci, L. (2016). Design and implementation of test equipment for SpaceFibre links: SpaceFibre, short paper. In: *2016 International SpaceWire Conference (SpaceWire)*, Yokohama, Japan.
9. Nannipieri, P., Leoni, A., & Fanucci, L. (2019). VHDL design of a SpaceFibre routing switch. *IEICE Transactions on Fundamentals of Electronics, Communications and Computer Sciences, E102A*(5), 729–731.
10. Nannipieri, P., Davalle, D., Fanucci, L., Leoni, A., & Jameux, D. (2019). Shine: Simulator for satellite on-board high-speed networks featuring SpaceFibre and SpaceWire protocols. *Aerospace, 6*(4), 43.
11. Scott, P., Mills, S., & Parkes, S. (2014). High speed test and development with the SpaceWire brick mk3. In: *2014 International SpaceWire Conference (SpaceWire)*, pp. 1–5. IEEE.
12. Steinbach, T., Korf, F., Bartols, F., & Schmidt, T. C. (2011). Performance analysis of time-triggered ether-networks using off-the-shelf-components. In: *2011 14th IEEE International Symposium on Object/Component/Service-Oriented Real-Time Distributed Computing Workshops*, Newport Beach, CA, USA, pp. 49–56.
13. Varga, A., & Hornig, R. (2008). An overview of the OMNeT++ simulation environment, cited By 1006. In: *Simutools08: 1st International Conference on Simulation Tools and techniques for Communications, Networks and Systems & Workshops*, Marseille, France, March, 2008.
14. Villafranca, A. G., et al. (2016). A new generation of SpaceFibre test and development equipment: SpaceFibre, short paper. In: *2016 International SpaceWire Conference (SpaceWire)*, pp. 1–4. IEEE.

15. Xilinx. Zcu102 evaluation board, ug1182 (v1.5). https://www.xilinx.com/support/documentation/boards_and_kits/zcu102/ug1182-zcu102-eval-bd.pdf
16. Xilinx. Virtex-6 family overview. https://www.xilinx.com/support/documentation/data_sheets/ds150.pdf
17. Yu, P., Koga, R., & George, J. (2008). Single event effects and total dose test results for TI TLK2711 transceiver. In: *2008 IEEE Radiation Effects Data Workshop*, pp. 69–75.

Chapter 6
Survey on Existing SpaceFibre-Based Solutions

6.1 Relevant FPGA Platforms

6.1.1 About the Effects of Space Radiations on Electronic Devices

It is known that the natural space radiation environment can lead to malfunctioning of electronic devices and also permanently damage the satellite electronic system [10]. For this reason, radiation tolerance is one of the most important constraints for space-qualified electronics. The principal effects of space radiations include the following:

- **Total Ionisation Dose** (TID): it is one of the most important requirements for a space-oriented electronic system. When a particle passes through a transistor, it can generate an electron–hole pair in its oxide [3]. The generated carrier induces a charge build-up, which leads to threshold voltage shift and leakage current in Complementary metal-oxide-semiconductor (CMOS) transistors and current gain degradation in bipolar transistors. The total accumulated dose depends on the orbit and time of exposure, and its effects can lead to a complete functional failure of the electronic system.
- **Single Event Effect** (SEE): it is caused by high-energy particles such as protons, neutrons and heavy ions that pass through sensitive regions of a microelectronic circuit, creating voltage glitches [1]. The term SEE is used to refer to different effects such as Single Event Upset (SEU) and Single Event Latch-up (SEL). In a SEU, the interaction between the incident radiation and the electronic system causes a temporary bit-flip that alters a datum but does not damage the electronic circuitry. A list of SEU effects comprehends, invalid generation of commands, spurious external memory accesses, deadlock in bus arbiters, data frame corruption, spontaneous triggering of resets and failure of internal reconfiguration commands. A SEL happens when high-energy particles determine a high-voltage

P. Nannipieri et al., *Next-Generation High-Speed Satellite Interconnect*,
https://doi.org/10.1007/978-3-030-77044-0_6

glitch that triggers a latch-up, a destructive event that can permanently damage the electronic circuit [2].

Considering that, it is necessary to properly design the electronic system to be as immune as possible from TID and SEE according to mission specifications. For example, in EO missions that target orbits between 100 km and 65,000 km from Earth, the presence of Van Allen belts makes TID the most hazardous threat for electronic circuitry. On the contrary, in deep space missions, SEE shall be considered the primary concern [11].

6.1.2 State-of-the-Art Rad-Hard FPGAs

Rad-hard FPGAs are devices that exploit the manufacturing process and design techniques to mitigate the effect of space radiations, limiting the adverse effects of TID and SEE. Indeed, the programmable logic of FPGAs allows realising custom architectures, which can be equipped with redundant structures to enhance reliability and robustness by design. For example, Error Detection and Correction (EDAC) techniques can be exploited to secure the transmission of data from/to Random Access Memories (RAMs) [12], and Triple Modular Redundancy (TMR) can mitigate SEU effects for combinatorial logic [13]. EDAC and TMR can also be implemented on non-rad-hard FPGAs, exploiting their programmable logic.

On the market, there are several vendors such as Microsemi, NanoXplore and Xilinx that produce rad-hard FPGAs. In the following, we will describe the features of the most relevant state-of-the-art rad-hard FPGAs:

- the *RTAX2000S/SL* is a 150 nm anti-fuse-based FPGA by Microsemi: anti-fuses are normally open circuits and form passive and low-impedance connections during the programming phase, and therefore they are one-time programmable devices. The RTAX2000S/SL features 21504 8-input multiplexers for implementing combinatorial logic and 10752 Flip-Flops (FFs). It has 64 4.5 kbits RAM blocks, for an overall on-chip memory of 288 kb. It does not embed Digital Signal Processors (DSPs) and transceivers. RTAX2000S/SL features TMR for protecting FFs and optional EDAC for RAM upset mitigation.
  The RTAX family can properly operate for a TID up to 300 krad and is SEL immune for a Linear Energy Transfer (LET) up to 117 MeV.cm2/mg.
- the *RTG4* is a 65 nm process flash-based FPGA by Microsemi. It includes 150k 4-input LUTs and FFs, 462 DSPs and two different kinds of RAM blocks: 64×18 uSRAMs and 18 kb large SRAMs for a total of 3.76 Mb on-chip memory. The RTG4 includes 24 high-speed transceivers with a maximum data rate of 3.125 Gbps each [14].
  RTG4 on-chip SRAM has a built-in EDAC and TMR protected FFs. The RTG4 family proved to sustain a TID up to 160 krad, and it is SEL immune for a LET up to 103 MeV.cm2/mg.

- the *PolarFire MPF500T* is a 28 nm flash-based FPGA by Microsemi. It includes 1924k 4-input LUTs, 1924k FFs and 1481 DSPs. Memory resources include 1520 20 kb LSRAM and 4440 64 × 12 uSRAM for an overall on-chip memory of 33 Mb. The PolarFire MPF500T has 24 transceivers that can operate up to 12.5 Gbps.
  For what concerns radiation tolerance, the PolarFire family maintains datasheet parameters for a TID up to 100 krad, and it is SEL immune for a LET up to 80 MeV.cm2/mg.
- the NanoeXplore *NG-Large* is a 65 nm SRAM-based FPGA [15], and it is the first European rad-hard FPGA. The NG-Large includes 137088 4-input LUTs, 129024 FFs, 384 DSPs and 192 RAM blocks of 48 kb for a total of 9 Mb of on-chip memory. The NG-Large features 24 transceivers operating up to 6,25Gbps. The Dual Interlocked storage Cell (DICE) memory latch has been used for Configuration Memory Cells and FFs. The NG-Large implements TMR, and it also foresees EDAC to protect BRAM blocks, but if enabled, the overall memory capacity is reduced to 6.75 Mb. The NG-Large is functionally unaffected by a TID up to 100 krad, and it is SEL immune for a LET up to 62 MeV.cm2/mg.
- the NanoeXplore *NG-Ultra* is 28 nm SRAM-based FPGA, and it is the largest rad-hard FPGA in the NanoeXplore catalogue. It includes 536928 4-input LUTs, 505344 FFs, 1344 DSPs and 672 RAM blocks of 48 kb that constitute the available on-chip memory (31.5 Mb). The NG-Ultra features 32 transceivers operating up to 12.50 Gbps. EDAC and TMR are adopted for protecting RAM blocks and FFs, respectively. If EDAC is enabled, the available on-chip memory decreases to 23.6 Mb.
  No details about TID and SEL immunity are provided by the vendor at the time of writing.
- the *Virtex-5QV* is a 65 nm SRAM-based FPGA developed by Xilinx. It includes 82 6-input LUTs, 82k FFs, 320 DSPs and 289 36 kb Block RAMs (BRAMs), for an overall on-chip memory of 10.7 Mb. BRAMs have embedded EDAC for mitigating SEU effects. The Virtex-5QV includes 18 4.25 Gbps transceivers.
  The Xilinx Virtex-5QV can sustain a TID up to 100 krad, and it is SEL immune for LET up to 100 MeV.cm2/mg.
- the Xilinx *Kintex UltraScale XQRKU060* is a 20 nm process SRAM-based FPGA, considered to be the next rad-hardened FPGA by Xilinx [16]. It includes 330k 6-input LUTs, 660k FFs, 2760 DSPs and an overall on-chip memory of 38Mb split into 36kb blocks. The XQRKU060 includes 36 13.2 Gbps transceivers.
  The Xilinx XQRKU060 proved to sustain a TID up to 100 krad, and it is SEL immune for a LET up to 79.2 MeV.cm2/mg [17].

The main features of the aforementioned state-of-the-art rad-hard FPGAs are summarised in Table 6.1.

Table 6.1 Summary of rad-hard FPGAs features

Vendor	Device	Technology [nm]	DSP	Memory [Mb]	TID (Krad)	SEL (LET)
Microsemi	RTAX2000	150	0	0.288	300	< 117 MeV.cm2/mg
Microsemi	RTG4	65	462	3.76	160	< 103 MeV.cm2/mg
Microsemi	PolarFire MPF500T	28	1481	33	100	< 80 MeV.cm2/mg
nanoeXplore	BRAVE Large	65	384	9.0	100	< 62 MeV.cm2/mg
nanoeXplore	BRAVE Ultra	28	1344	31.5	–	–
Xilinx	Virtex-5QV	65	320	10.7	100	< 100 MeV.cm2/mg
Xilinx	Kintex XQRKU060	20	2760	38.0	100	< 79.2 MeV.cm2/mg

6.2 SpaceFibre CoDecs

In the literature, there are several SpaceFibre CoDec IP core implementations, and this section presents an exhaustive report of the solutions developed so far.

STAR-Dundee, the company responsible for the standard publication, implemented its version of the SpaceFibre IP (both single-lane and Multi-Lane versions) onto various space-grade FPGAs such as the Microsemi RTG4, Microsemi RT-PolarFire, Xilinx Virtex-5QV and Xilinx Kintex UltraScale KU060 [18]. An alternative implementation on-board the Microsemi RTG4 is presented in [5], including more details about memory occupation. Finally, additional data for the Multi-Lane version of the IP core for a different combination of VCs and a number of lanes are presented in [19]. Also, ESA has the right to sublicense the Cobham Gaisler single-lane SpaceFibre IP within its IP core portfolio [20] for Microsemi RTG4, Microsemi RTAX2000 and Xilinx Virtex-5QV rad-hard FPGAs. Finally, IngeniArs implements its single-lane and the Multi-Lane versions of the SpaceFibre CoDec IP core on-board several rad-hard FPGAs, also including power consumption results for the Microsemi RTG4. Table 6.2 presents a complete recap of the SpaceFibre CoDec IP core available in the literature.

In general, the higher is the number of VCs and lanes, the higher is the number of resources required for the implementation of a SpaceFibre IP core. A direct comparison between the IP cores proposed by STAR-Dundee, IngeniArs and Cobham Gaisler can be presented for the single-lane 4 VCs implementation on-board an RTG4, as it is the most commonly referenced configuration. From the numbers available, we can observe similar resource utilisation in all the FPGAs, with small variations that can be linked with different optimisations. We can observe that the IngeniArs implementation reaches the maximum available bandwidth. Compared to the others, it employs a similar number of logic elements (even if the ratio between LUTs and FFs is inverted) but manages to reduce the total memory usage by 25% with respect to the Cobham Gaisler implementation and 56% with respect to the STAR-Dundee one.

Table 6.2 Summary of SpaceiFibre IP core implementation on rad-hard FPGAs

	FPGA	Lanes	VCs	LUT %	FF %	Memory [kb]	Power [mW]
STAR-Dundee [18]	RTG4	1	2	2.5%	1.8%	–	–
		4	2	9.0%	6.8%	–	–
STAR-Dundee [18]	RT-PolarFire	1	2	0.7%	0.6%	–	–
		4	2	2.8%	2.1%	–	–
STAR-Dundee [18]	Virtex-5QV	1	2	3.4%	1.4%	–	–
		4	2	9.6%	11.7%	–	–
STAR-Dundee [18]	Kintex UltraScale KU060	1	2	0.6%	0.4%	–	–
		4	2	2.0%	1.4%	–	–
STAR-Dundee [5]	RTG4	1	4	4.8%	2.1%	324	–
STAR-Dundee [19]	RTG4	2	1	4.3%	3.5%	144	–
		2	2	4.8%	4.0%	216	–
		3	2	5.9%	4.8%	216	–
STAR-Dundee [19]	Virtex-5QV	2	1	4.7%	4.8%	144	–
		2	2	5.5%	5.3%	216	–
		3	2	6.6%	6.4%	216	–
IngeniArs [8]	RTG4	1	1	3.1%	1.5%	134	331.7
		1	2	3.8%	1.9%	152	349.3
		1	4	4.7%	2.5%	188	369.8
		1	8	6.6%	3.8%	260	424.5
IngeniArs [8]	Virtex-5QV	1	1	3.6%	2.1%	252	–
		1	2	4.5%	2.7%	432	–
		1	4	6.2%	3.7%	576	–

(continued)

Table 6.2 (continued)

	FPGA	Lanes	VCs	LUT %	FF %	Memory [kb]	Power [mW]
IngeniArs [8]	RTAX-2000	1	1	28.6%	26.2%	54	–
		1	2	35.4%	33.4%	72	
		1	4	45.9%	44.7%	108	–
IngeniArs [7]	RTG4	2	1	7.4%	3.8%	232.9	483
		2	2	8.7%	4.5%	235.1	488
		2	4	10.9%	6.2%	239.6	531
		3	1	11.1%	5.3%	275.6	574
		3	2	12.9%	6.1%	279.0	614
		3	3	16.0%	7.6%	285.8	679
		4	1	14.1%	6.8%	318.4	681
		4	2	16.4%	7.8%	322.9	729
		4	4	20.7%	9.7%	331.9	760
IngeniArs [7]	Kintex UltraScale KU060	2	1	2.7%	2.1%	288	–
		2	2	3.1%	2.3%	378	–
		2	4	3.6%	2.5%	558	–
		3	1	4.0%	3.1%	360	–
		3	2	4.5%	3.3%	594	–
		3	4	5.3%	3.6%	738	–
		4	1	5.3%	4.1%	432	–
		4	2	6.0%	4.1%	594	–
		4	4	7.1%	4.4%	918	–
Gaisler [20]	RTG4	1	4	3.8%	1.6%	180	–
Gaisler [20]	RTAX-2000	1	2	27.0%	19.0%	200	–
Gaisler [20]	Virtex-5QV	1	4	4.0%	2.0%	162	–

Table 6.3 IngeniARS SpFi CoDec IP 1 VC Power Consumption on RTG4 with BER = 10^{-5}

BER	Static Pwr [mW]	Dynamic Pwr [mW]	Total Pwr [mW]
0	183.3	148.4	331.7
10^{-5}	183.3	149.7	333.0

The IngeniArs implementation is the only one that includes a power consumption analysis. As indicated in [7, 8], the power consumption has been measured for the Microsemi RTG4 board through post-implementation simulations. Results have been obtained with 200 μs simulations corresponding to the transmission of about 520 kBit of data divided in 36-bit data words, interleaved with 1 broadcast message for each 1K data words. The operating frequency has been set to 62.5 MHz. The data in Table 6.2 do not take into account SerDes power consumption, which can be estimated to be 285 mW (the value obtained with the official Microsemi RTG4 power estimator tool). The computed dynamic power is to be considered a good estimation since the testbench stimulates the device under test in its maximum power dissipation scenario. The analysis proposed by IngeniArs also measures the power consumption on a faulty link with a Bit Error Rate (BER) (the worst value tolerated in the SpFi standard [6]) for the single VC configuration [8] of its IP core. This BER value represents a worst-case scenario since realistic BER values are in the order of 10^{-11}–10^{-13} [21]. Table 6.3 reports both dynamic power, determined by the circuit switching activity, and the static power that characterised each FPGA. Data indicates that the change in the overall power consumption is marginally affected by system BER, with an increase of just 1.3 mW. This result appears to be reasonable because considered the BER of 10^{-5} used in the test, the portion of the circuit dedicated to the retry mechanism is stimulated only one bit out of 10^{-5}, which means roughly that 1 word out of little less than 3000 is corrupted, triggering the activation of the retry mechanism. Moreover, when the retry mechanism is activated, other parts of the circuit are deactivated (e.g. the VCs); therefore, dynamic power is saved somewhere else. However, this does not take into account that, depending on the BER, the time necessary to deliver a packet is higher; therefore, also the total energy necessary to deliver every single bit will be higher. However, this value is not directly affected by the increased power consumption of the circuit, but only by the longer transmission time required.

6.2.1 *Multi-Lane Recovery Time*

In this section, we present the performance of the SpaceFibre standard in terms of recovery time after a lane failure, as described in the available literature. The overall recovery time consists of two contributes:

- the *Multi-Lane layer recovery time*, which is the time required by the Multi-Lane layer to realign data words after a lane failure.

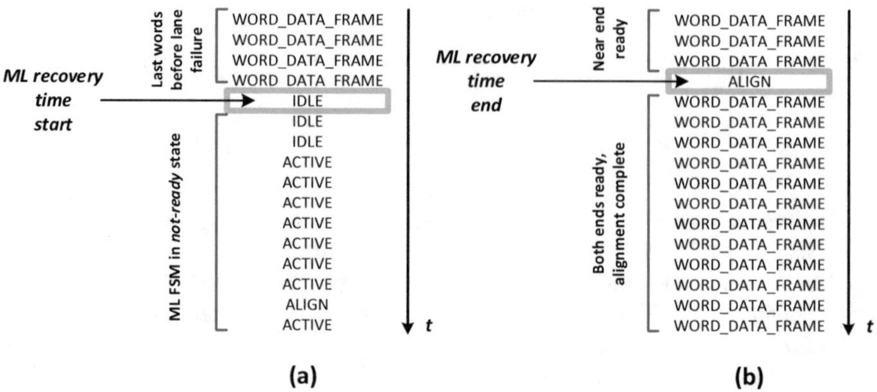

Fig. 6.1 Multi-Lane recovery time measured using SpaceART® [7]

- the *Data-Link layer recovery time*, which is the time that the Data-Link layer needs for retransmitting the frame that it was transmitting at the time the lane failed.

That Data-Link layer can start to elaborate and transmit a new data frame only when the recovery time is completed.

The work presented in [7] includes a detailed analysis of the recovery time for the IngeniArs IP core. The test aims at simulating the traffic between 2 SpaceFibre end-nodes, continuously feeding two IP cores with new random data that are then sent to the other end of the link. The SpaceART® equipment tool was employed to generate/consume and monitor the traffic over the SpFi link [9, 22, 23]. The Multi-Lane recovery time starts when the Lane layer begins to send IDLEs, which are special control words transmitted when no valid data are available from the upper layers, and it ends when the last Multi-Lane control word is sent, and the Multi-Lane layer enters the *Both ends ready* state, as specified by the SpFi standard [6]. It is possible to monitor the link traffic and identify the exact moment in which the Lane layer starts to send IDLEs exploiting SpaceART® features, as shown in Fig. 6.1a. Then, the Multi-Lane layer starts to send ACTIVE and ALIGN control words to the far-end of the link to perform lane alignment. Finally, when the alignment process is complete, and the ML-FSM sends the last ALIGN control word entering the *Both ends ready* state, the ML recovery times can be considered complete, as indicated in Fig. 6.1b.

The total recovery time also includes the contribution of the Data-Link recovery time. Indeed, the data frame that is transmitted when a lane fails is inevitably corrupted and shall be retransmitted. Before the retry operation can start, some data words pipelined in the transmitting side at the time of the disconnection re-connection may be received. Also, the Data-Link layer could start to send the very next frame before serving the retry request. As shown in Sect. 2.5, the Data-Link layer provides an FDIR system to re-send a data frame that contains an error, and

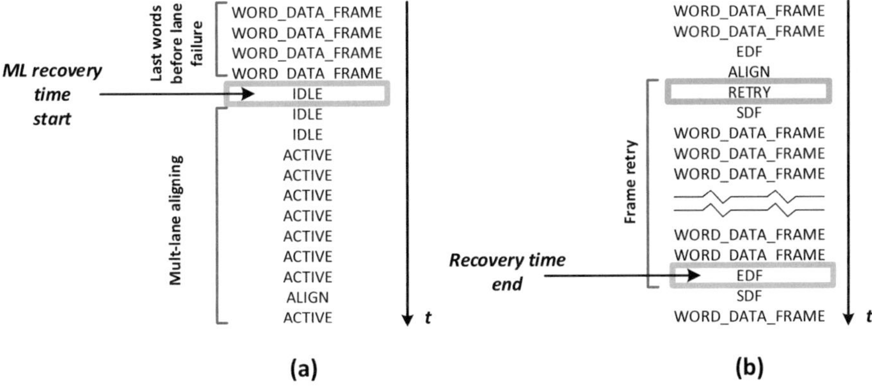

Fig. 6.2 Total recovery time, including Data-Link and Multi-Lane layers recovery time, measured using SpaceART® [7]

Table 6.4 Recovery time for Multi-Lane SpaceFibre IP core

Case	VCs	Avg. ML rec. time [Clk cycles]	Avg. DL rec. time [Clk cycles]	Tot. [Clk cycles]	Tot. [μs]
(1) (1− > 2) [7]	1	77.4	198.58	276	4.42
(2) (1− > 2) [7]	2	81	244.25	323.25	5.17
(3) (2− > 1) [7]	1	79	272.62	351.62	5.26
(4) (2− > 1) [7]	2	79	314.48	393.48	6.30

once the Multi-Lane alignment is completed, it re-sends the data frame through the Retry Buffer. In Fig. 6.2a, the Multi-Lane layer starts the alignment process. Figure 6.2b shows the moment when the retry process starts: the Data-Link layer transmits the RETRY control word to signal that the next data frame is part of a retry process. When the Data-Link receives an End of Data Frame (EDF), a complete frame has been received and the recovery time ends. Data measured using SpaceART® are shown in Table 6.4, which include the Data-Link layer recovery time, the Multi-Lane recovery time and the total recovery time for four different test cases: cases (1) and (2) start with one active lane and one inactive lane; the second lane is then reactivated (1− > 2). Cases (3) and (4) begin with two active lanes; one lane is then deactivated (2− > 1). Cases (1) and (2) and cases (3) and (4) differ for the number of VCs considered. To avoid multiple connections and re-connections as a consequence of the manual disconnection of the cable connecting the two IPs, a mechanism to arbitrarily modify the status of a lane from "*active*" to "*inactive*" and vice versa has been implemented. Once the Multi-Lane layer recognises that a lane changes its status, it immediately starts the alignment process.

The difference between the total recovery time of cases (1) and (2) and cases (3) and (4) can be attributed to the different data rate of the two systems after the re-connection/disconnection process. Indeed in cases (1) and (2), the IP can exploit the maximum achievable data rate since all the lanes are active. The number of active lanes does not influence the ML recovery time, which is approximately the same for all the cases, but it determines the DL recovery time.

Cases (2) and (4) include 2 VCs that try to communicate over the physical link. The total recovery time appears to be higher than the cases featuring 1 VC because the Data-Link QoS cannot firstly schedule the VC that was transmitting at the time of the lane re-connection/disconnection, but it can schedule other VCs before serving the retry request. In all the cases considered, the recovery time is in the order of a few μs, assuming a system frequency of 62.5 MHz.

6.3 Routing Switches

The final specification of the network layer has been inserted in the SpaceFibre short before the final publishing. The release of the SpaceFibre standard itself is indeed quite recent (May 2019) at the time of writing. However, the first draft of the standard, which helped the SpaceFibre community to start building CoDecs and other Ips, is dated earlier than 2013. At that time, the network layer was only drafted within the standard, and the first more complete version is dated 2017, with major changes also occurred later. For this reason, together with the fact that currently, no SpaceFibre point-to-point links have already flight heritage, companies do not see the SpaceFibre router as a fundamental building block in the forthcoming year. It is more probably something that will get a larger adoption rate in the second part of the decade (2025–2030), once the point-to-point technology will have more heritage. In this context, the only two available routers belong to the two companies more involved in the SpaceFibre standardisation effort: IngeniArs and STARDundee. In the following, we will try to present their SpaceFibre routers and, where possible, provide a fair comparison among them.

6.3.1 IngeniArs Router

The routing switch largely described in Sect. 3.3.1 and presented also in [4] has been synthesised for different technologies, considering both consumer-grade FPGAs commonly used for laboratory and test equipment (e.g. EGSEs and link analysers) and radiation-hardened FPGAs, suitable to host a SpaceFibre routing switch to be used in a spacecraft. The presented results include all the logic embedded in the switching block; the SpaceFibre and SpaceWire ports are not included and their complexity can be derived from the results presented in Sect. 6.2. The Xilinx Virtex-6 XC6VLX240T and the Xilinx Zynq XC7Z045 have been selected as commercial products and the Microsemi RTG4 as the space-grade FPGA. The number of ports and the number of VCs are the factors mostly affecting the design complexity, and therefore we performed several syntheses runs varying them, to study the impact

Table 6.5 Synthesis results for Microsemi RTG4 [4]

Set-up	Triple redundancy reg.	4-Input LUT	1,5 Kb RAM Block	Max. Freq.
4 ports, 4 VCs	4.9%	12.8%	6.7%	83.1 MHz
4 ports, 8 VCs	10.9%	37.8%	26.1%	74.5 MHz
8 ports, 4 VCs	11.1%	39.2%	16.6%	72.2 MHz
8 ports, 8 VCs	26.6%	132.5%	28.5%	51.4 MHz
4 ports, 4 VCs, no multicast	4.3%	11.7%	5.7%	92.4 MHz
8 ports, 8 VCs, no multicast	24.0%	120.4%	26.1%	65.8 MHz

Table 6.6 Synthesis results for Xilinx Virtex-6 [4]

Set-up	Single bit reg.	6-Input LUT	18 Kb RAM blocks	Max. Freq.
4 ports, 4 VCs	2.1%	9.8%	0.5%	187.1 MHz
4 ports, 8 VCs	5.3%	24.3%	0.5%	163.5 MHz
8 ports, 4 VCs	6.2%	26.1%	1.0%	164.2 MHz
8 ports, 8 VCs	14.9%	78.8%	1.0%	140.4 MHz

Table 6.7 Synthesis results for Xilinx Zynq [4]

Set-up	Single bit reg.	6-Input LUT	18 Kb RAM blocks	Max. Freq.
4 ports, 4 VCs	1.7%	5.9%	0.18%	205.3 MHz
4 ports, 8 VCs	3.8%	16.4%	0.18%	188.8 MHz
8 ports, 4 VCs	3.9%	15.9%	0.36%	192.1 MHz
8 ports, 8 VCs	9.4%	50.3%	0.36%	161.4 MHz

of these parameters. These preliminary results can then be used by future system adopters as a valid estimation of the required resources. The tools used to synthesise and implement are Xilinx ISE 14.6 and Xilinx Vivado 2016.4, respectively, for the Virtex-6 and the Zynq and Synplify Pro 14 for the RTG4. An idea of the complexity of the multicast mechanism can be obtained from the two extra synthesis results shown in Table 6.5, where the resource utilisation without multicast is presented for two configurations.

Tables 6.5, 6.6 and 6.7 report synthesis results for the three FPGAs together with the maximum estimated clock frequency. As expected, by increasing the number of ports and/or their VCs, we obtain a more complex design: e.g. using 8 ports with 8 VCs each leads to the impossibility to fit the design in the RTG4. This is not considered to be a problem, as ASIC solutions are usually preferred in-flight hardware for the interconnection backbones. The routing switch fits instead in the commercial solutions even with a larger number of ports and VCs, where an ASIC solution is usually adopted. Considering the internal 36-bit wide data path, the frequency goal to reach a link rate of 2.5Gbps is 62.5MHz, which has been achieved for all the presented configurations.

Table 6.8 4-Port 4-VC STAR-Dundee Router and SpFi ports synthesis results for Microsemi RTG4 [4]

Set-up	Triple redundancy reg.	4-Input LUT	1,5 Kb RAM block
Static VN	11.7%	20.4%	14.4%
Dynamic VN	22.6%	38.4%	22.0%

6.3.2 STAR-Dundee Router

STAR-Dundee presented its version of the SpaceFibre router in 2017 [24] and [25], with the name SUNRISE. Unfortunately, the amount of information given on this IP is very limited. It is indicated that it has a configurable number of ports and VCs and that it includes an RMAP configuration port. This router introduces a new feature: the user can decide at synthesis time whether the Virtual Network shall be static or dynamic (e.g. if the Virtual Network can be changed with an RMAP packet or not). This optimisation led to potential resource-saving, shown in Table 6.8.

The data presented in [25] and reported in Table 6.8 are not referred to a specific configuration of the router. However, these data refer to a routing switch with SpaceFibre/SpaceWire ports. Therefore, any specific comparison needs to be postponed waiting for more specific data.

6.4 Electrical Ground Segment Equipment

Another area in which the SpaceFibre technology will be employed for sure is the *EGSE* field of application. It is more a self-arisen consequential demand rather than a technical exercise for its own sake. The need to implement an EGSE SpFi emerges at the exact moment that the intended use of this technology is decided for the upcoming mission. If a spacecraft of a sort is to be designed and manufactured employing SpFi technology as on-board communication link technology, then extensive and effective support testing equipment will be required to verify its compliance and proper functioning. Indeed, a spacecraft shall be tested both as a single component level and as a whole system synergetically. EGSE systems belong to the ground segment of a space mission (e.g. all support equipment and services necessary for the fulfilment of the mission itself, carried out by the space segment). *Electrical Ground Segment Equipment* occupies a clearly defined part within a space mission. EGSE testing solutions have a centre role especially in the *detailed definition phase* and the *qualification and production phase* to support development, manufacturing and integration of spacecraft on-board devices and sub-systems. These pieces of equipment are therefore necessarily as unique as the satellites they are intended to support. The reader must keep in mind the special nature and uniqueness of each spacecraft, at least in its payload that is tightly bonded to the mission scope and purposes. This may seem in the first instance to be in blatant contrast with the purpose of a standardisation process to encourage reuse and interoperability, from which the definition of a protocol such as SpaceFibre originates and is fulfilled, but it is not contradictory at all. Indeed, the standardisation effort seeks—if this concept has not yet been made clear—to empower the final users to design their own mission-specific vessels by providing basic functional elements to be combined. Figure 6.3 shows a generic architecture of an EGSE system and its connections. As testing equipment, it generates an output to stimulate a given device of interest and then collects as input its response, performing meaningful and qualitative analysis. This answers exhaustively the question of which essential features and requirements EGSE systems must address: observability and controllability. Figure 6.3 is also particularly meaningful in pointing out how the EGSE is connected to other systems: an EGSE system finds its rationale in this specific connection. Designing EGSE test equipment for a given protocol poses additional challenges to provide the desired observability and controllability beyond the sole communication with the DUT. Let us consider the case, which we may be particularly interested in, of an EGSE system for an on-board data-handling link technology such as SpaceFibre: a system could not be considered a suitable candidate as a SpFi EGSE system solely because it is equipped with the correct communication interface (i.e. a SpFi CoDec). An effective and efficient EGSE system must also provide the user with all the tools for the needed observability and controllability over the link of interest, and in particular:

- Observability refers to the ability to make an internal state (a variable, a signal, etc.) about the system of interest measurable, or inferable at least.
- Controllability refers to the ability to apply a desired stimulus or force a required internal state on the system of interest.

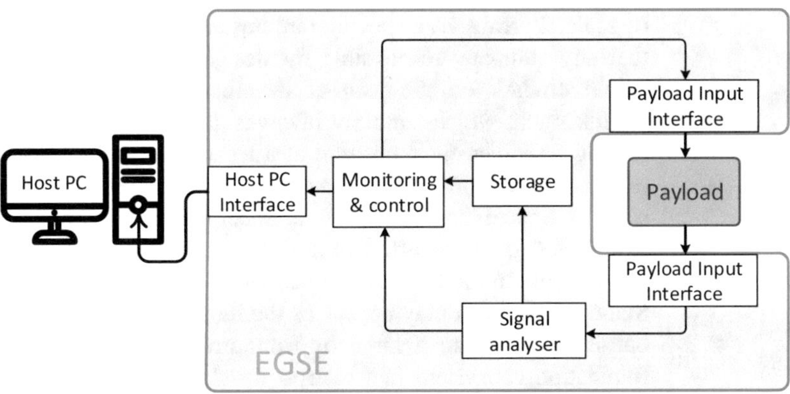

Fig. 6.3 Generic EGSE block diagram

6.4.1 State-of-the-Art EGSE Solutions for SpaceFibre Technology

The EGSE solutions available on the market can be roughly divided into three groups by their scope of implementation:

- *Custom-designed solutions*: test equipment systems explicitly and limitedly designed, from hardware to software level, for a single mission or a single class of missions. These types of solutions are offered indirectly in the marketplace as design services.
- *Proprietary solutions*: test equipment systems built on a dedicated hardware platform developed by a company and that can be used effectively for any mission, with minor customisations eventually.
- *Standard-platform solutions*: test equipment systems that implement EGSE functionality on standard general-purpose COTS hardware, usually modular and thus expandable, within a broader testing ecosystem (e.g. a platform).

Each implementation approach has its strengths that may lead to the former being preferable to the latter upon the very specific needs of the given mission, the particular system (or sub-system) to be tested, and the context to which the equipment is meant to be integrated. Indeed, the survey presented in this paragraph outlines the following possible key figures from the research and analysis carried out on the market and its stakeholders that may impact the final decision on the type of equipment to be adopted:

Affordability takes into account the economic investment needed to set up the EGSE equipment as a whole testing environment solution. Common sense is that a custom solution has a higher cost of manufacturing because small volumes do not benefit from economies of scale. Barring very specific circumstances, it is very unlikely that any company responsible for designing an EGSE solution would embark on the path of developing a custom solution from scratch, which certainly involves a lot of risks rather than relying on either customisation of a legacy proprietary solutions or standard platform. Often though, for new emerging technology such as SpaceFibre, there are no specifically developed EGSE solutions yet; so, the effort to design a custom solution becomes the momentum for the establishment of a new proprietary solution. Standard platform may appear as the most cost-effective option, but it may require a large up-front investment to set up the foundation ecosystem in which to develop EGSE functionality. For this reason, standard-platform solutions are preferred by prime customers (who value other figures more), while proprietary solutions manage to gain market share among those sub-contractors in the mission hierarchy.

Reliability involves both the reliability of the single hardware component in the EGSE equipment and the reliability of the providing manufacturer/distributor. Standard-platform solutions mitigate the risk of letting the end-user uncovered and without support, either because a popular industry standard is supported by a joint group interest of several manufacturers or because the individual products are intended for several end markets, custom design instead. Reliability is the primary key figure for prime contractors in the space industry, much higher on the scale of their priorities than the mere economic cost.

Integration addresses the requisite or quantifies the effort, to incorporate the EGSE tool into an existing testing environment, or the EGSE tool ability to establish such an environment that can incorporate other components and work with them in unison; the integration effort also should take into account the time it requires for new users to master the tool and how much the new instrument changes or differs from the typical, expected or previously existing workflow.

Scalability concerns how well a testing solution can be employed on different granularity (e.g. if the instrument is suitable to test every single component) as well as sub-systems or the whole system. This key feature is valuable mostly to prime contractors and large system integrators because having a single tool for all needs streamlines the maintenance of equipment and environments regardless of the kind of targets to be tested, especially, simplify company procedures for quality controls.

Expandability is intended to provide an understanding of the testing equipment that is potential to add new features and support new hardware or software components. Standard platform-based solutions, for instance, can incorporate new features by installing additional modules. On the other side instead, custom or proprietary solutions are more monolithic and an eventual expansion requires a real customisation operation (see the next decription).

Customisation refers to the likeliness of the testing equipment provider to modify its reference design accordingly to the customer's need (e.g. to support a feature not foreseen or address a very specific requirement). Custom and proprietary solutions are, of course, more inclined to offer this possibility to their customers; standard-platform manufacturers usually are not yet sometimes can offer customisation services by forwarding the request to any of their business partners.

SpaceFibre is an emerging technology intended, yet not limited, for the niche of space-based applications. There are only three players in the market offering EGSE solutions for SpaceFibre at the time this manual was written. The following

Table 6.9 Brief comparison of the presented EGSE equipment

Feature	STAR-Dundee StarFire MK3	TeleTel Quad-SpFi for iSAFT	IngeniArs SpFi PXI
Implementation	Custom-designed	Proprietary	Standard platform
SpFi IFs	2	4	4 or 8
SpW IFs	2	0	4
Host IF	USB 3.0	PCIe	PXIe
Triggers (in/out)	1 / 1	0	as per PXI bus
SpFi min rate	1.88 Gbps	1 Gbps	1 Gbps
SpFi max rate	3.2 Gbps	3.2 Gbps	6.25 Gbps
SpFi max VCs	8	16	8
Hardware packet generation	Yes	Yes	Yes
Hardware packet consumption	Yes	Yes	Yes
Host packet generation	Yes	Yes	Yes
Host packet consumption	Yes	Yes	Yes
Host packet recording	Yes	Yes	Yes
Low-level protocol trace memory	Yes	Yes	Yes
Unobtrusive monitoring	Yes	Yes	Yes
Packet sniffing	Yes	Yes	Yes
User experience	Proprietary GUI Applications	C++/C# API	LabVIEW Application and LabVIEW API

paragraphs briefly present some EGSE possible solutions and Table 6.9 compares them in terms of offered features.

6.4.2 STAR-Dundee StarFire MK3

STAR-Dundee is a spin-off company of the University of Dundee (Scotland, UK) founded by Steve Parkes following joint efforts with ESA to define the SpaceWire standard. STAR-Dundee also played a prime role in the SpaceFibre Working Group for the definition and promotion of the SpaceFibre link technology. Given its background, STAR-Dundee represents a highly creditable subject for both SpaceWire and SpaceFirbe standards. STAR-Dundee only implements custom EGSE solutions for SpaceFibre, although it had previously developed compatible EGSE solutions as well for the standard LabVIEW platform for SpaceWire technology. STAR-Fire

Mk3 is a platform developed to support the testing and evaluation of SpaceFibre technology. It features SpaceWire to SpaceFibre bridging, link analysis capabilities, internal SpaceWire and SpaceFibre pattern generators and checkers for multiple virtual channels.

Offered Interfaces

- 2 SpFi compatible interfaces, at 1.88 or 3.2 Gbps;
- 2 SpW compatible interfaces, running up to 400 Mbps;
- 1 USB 3.0, used for the host-PC communication;
- 1 SMB interface, used as an external trigger input;
- 1 SMB interface, used as an external trigger output.

Offered Features

- SpFi/SpW bridging;
- 8 virtual channels per SpaceFibre interface;
- Error injection capabilities;
- Broadcast message transmission/reception;
- Tracing memory to implement link analyser capabilities;
- Interface diagnostics while sending and receiving SpFi packets;
- Unobtrusive monitoring and sniffing of a SpFi link;
- SpFi and SpW on-board packet generation and checking;
- SpFi and SpW packets generation and consuming on host-PC;
- Host-PC communication based on USB 3.0;
- Multiple dedicated and focused GUI applications for link control, link statistics and link analysis.

6.4.3 TeleTel Quad SpaceFibre PCIe IF Card for iSAFT

TeleTel is a company located in Greece with more than 25 years of experience providing hardware and software solutions for communication systems in the space, defence and aeronautics fields of application. Their product portfolio covers the majority of the interfaces used in space missions of agencies from all over the world, including the proprietary standard TTEthernet. TeleTel only develops custom EGSE solutions for SpaceFibre technology as additional modules for their proprietary iSAFT platform. The Quad SpaceFibre IF card is a module for the iSAFT line-up, a TeleTel proprietary platform for simulation, validation and monitoring of many interface protocols used for spacecraft on-board communications. TeleTel proposed that SpaceFibre solution is based on the industry-proven SpaceFibre CoDec validated in ESA representative SpaceFibre testbenches. It is suitable either for EGSE purpose or for SpFi device prototyping.

Offered Interfaces

- 4 SpFi compatible interfaces, at 1, 1.25, 2, 2.25 or 3.2 Gbps;
- PCIe remote host interface port.

Offered Features

- Up to 16 virtual channels per SpaceFibre interface;
- SpFi data reception and packet truncation support;
- SpFi broadcast message transmission/reception;
- SpFi data/broadcast reception time-stamping;
- SpFi statistics support for transmission, reception or broadcast packets;
- host-PC communications via PCIe interface;
- C#/C++ Application Programming Interface (API).

6.4.4 IngeniArs SpFi PXI Analyser

IngeniArs is a spin-off company of the Pisa University (IT). This start-up also stemmed from the efforts in defining and promoting the SpaceFibre standard, by participating in the work of the definition committee. IngeniArs develops and owns its proprietary implementations for both SpaceWire and SpaceFibre technology, and its authority in the field is bolstered by the many valuable participation in missions for several space agencies. IngeniArs develops SpaceFibre and SpaceWire EGSE solutions either as custom-designed hardware and software or as based on general-purpose FPGA-based PXI peripheral modules for the LabVIEW standard platform. In Sect. 3.4, the SpaceART® unit has already been presented. In addition to their custom-developed solution for SpaceFibre testing equipment, IngeniArs also offers in its own portfolio a standard-platform solution of equivalent functionalities developed on general-purpose National Instruments FPGA PXI peripheral modules; SpFi and SpW protocol-compliant connections (e.g. micro-D9 or eSATA ports) are offered via additional adapter modules though.

Offered Interfaces

- 4 or 8 SpFi compatible interfaces (via connector adapter), up to 6.25 Gbps;
- 4 SpW compatible interfaces (via connector adapter), up to 200 Mbps;
- PXI interface for host-PC communications.

Offered Features

- SpFi/SpW bridging;
- Up to 8 virtual channels per SpaceFibre interface;
- Error injection capabilities;
- Tracing memory to implement link analyser capabilities;
- SpFi and SpW on-board packets generation and checking;
- SpFi and SpW packets generation and consuming on host-PC;

- Natively integrated in National Instruments LabVIEW Environment;
- LabVIEW Application Programming Interface (API).
- LabVIEW GUI Companion Application.

References

1. Normand, E. (1996). Single-event effects in avionics. *IEEE Transactions on nuclear science, 43*(2), 461–474.
2. Bruguier, G., & Palau, J. M. (1996). Single particle-induced latchup. *IEEE Transactions on Nuclear Science, 43*(2), 522–532.
3. Xu, Y. N., Xu, G. B., Wang, H. B., Chen, L., Bi, J. S., & Liu, M. (2017). Total ionization dose effects on charge-trapping memory with al 2 o 3/hfo 2/al 2 o 3 trilayer structure. *IEEE Transactions on Nuclear Science, 65*(1), 200–205.
4. Nannipieri, P., Leoni, A., & Fanucci, L. (2019). VHDL design of a SpaceFibre routing switch. *IEICE Transactions on Fundamentals of Electronics, Communications and Computer Sciences, E102A*(5), 729–731.
5. McClements, C., McLaren, D., Youssef, B., Ali, M. S., Florit, A. F., Parkes, S., & Villafranca, A. G. (2016). SpaceWire and SpaceFibre on the microsemi RTG4 FPGA. In: *2016 IEEE Aerospace Conference*, Big Sky, MT, USA, 2016-June.
6. European Cooperation for Space Standardisation. (2019). *SpaceFibre – Very high-speed serial link, ECSS-E-ST-50-11C*. European Cooperation for Space Standardisation.
7. Nannipieri, P., Marino, A., Fanucci, L., Dinelli, G., & Dello Sterpaio, L. (2020). The very high-speed SpaceFibre multi-lane codec: Implementation and experimental performance evaluation. *Acta Astronautica, 179*, 462–470.
8. Dinelli, G., Marino, A., Dello Sterpaio, L., Leoni, A., Fanucci, L., Nannipieri, P., & Davalle, D. (2020). A serial high-speed satellite communication codec: design and implementation of a SpaceFibre interface. *Acta Astronautica, 169*, 206–215.
9. Leoni, A., Dello Sterpaio, L., Davalle, D., & Fanucci, L. (2016). Design and implementation of test equipment for SpaceFibre links: Spacefibre, short paper. In: *2016 International SpaceWire Conference (SpaceWire)*, Yokohama, Japan.
10. Winokur, P. S., Fleetwood, D. M., & Dodd, P. E. (2000). An overview of radiation effects on electronics in the space telecommunications environment. *Microelectronics Reliability, 40*(1), 17–26.
11. European Cooperation for Space Standardization (ECSS). (2016). *Techniques for radiation effects mitigation in ASICs and FPGAs handbook*. ECSS.
12. Bentoutou, Y. (2012). A real time EDAC system for applications onboard earth observation small satellites. *IEEE Transactions on Aerospace and Electronic Systems, 48*(1), 648–657.
13. Ramos, J., Samudrala, P. K., & Katkoori, S. (2004). Selective triple modular redundancy (STMR) based single-event upset (SEU) tolerant synthesis for FPGAs. *IEEE transactions on Nuclear Science, 51*(5), 2957–2969.
14. Microsemi. Microsemi rtg4 datasheet. https://www.microsemi.com/product-directory/rad-tolerant-fpgas/3576-rtg4
15. NanoXlore. Nanoxlore NG-Large datasheet. Available online: https://www.nanoxplore.com/uploads/NanoXplore_NG-LARGE_Datasheet_v1.0.pdf
16. Xilinx. Xilinx Kintex usage for space application. Available online: https://www.xilinx.com/products/silicon-devices/fpga/rt-kintex-ultrascale.html
17. Xilinx. Xilinx Kintex ultrascale datasheet. https://www.xilinx.com/support/documentation/data_sheets/ds890-ultrascale-overview.pdf
18. Casas, M. F., Parkes, S., Villafranca, A. G., Florit, A. F., & McClements, C. (2020). Spacefibre for FPGA: IPs and radiation test results. In: *29th Annual Single Event Effects (SEE) Symposium coupled with the Military and Aerospace Programmable Logic Devices (MAPLD) Workshop)*.

19. Villafranca, A. G., Parkes, S., & Florit, A. F. (2016). Spacefibre multi-lane: Spacefibre, long paper, In: *2016 International SpaceWire Conference (SpaceWire)*, pp. 1–8, http://dx.doi.org/10.1109/SpaceWire.2016.7771647
20. ESA. ESA HDL IP cores portfolio overview. Available online: https://www.esa.int/Enabling_Support/Space_Engineering_Technology/Microelectronics/ESA_HDL_IP_Cores_Portfolio_Overview
21. Yu, P., George, J., & Koga, R. (2008). Single event effects and total dose test results for TI TLK2711 transceiver. In: *2008 IEEE Radiation Effects Data Workshop*, pp. 69–75.
22. Nannipieri, P., Marino, A., Dello Sterpaio, L., & Fanucci, L. (2019). Design of a spacewire/spacefibre EGSE system based on PXI industry standard. In: *2019 IEEE International Workshop on Metrology for AeroSpace*, pp. 1–5. IEEE.
23. Marino, A., Nannipieri, P., Dinelli, G., Davalle, D., Dello Sterpaio, L., & Fanucci, L. (2019). A complete egse solution for the spacewire and spacefibre protocol based on the PXI industry standard. *Sensors, 19*(22), 5013.
24. Florit, A. F., Villafranca, A. G., McClements, C., Parkes, S., & McLaren, D. (2017). SpaceFibre network and routing switch. In: *2017 IEEE Aerospace Conference*, Big Sky, MT, USA.
25. Villafranca, A. G., Parkes, S., Florit, A. F., & McClements, C. (2018). Spacefibre interface and routing switch IP cores. In: *2018 International SpaceWire Conference*.

Chapter 7
Conclusions

In Chap. 1, a brief introduction on satellite on-board data-handling sub-system is given. The main communication link technologies currently available on the market (SpaceWire, TTEthernet, WizardLink, RapidIO and SpaceFibre) are presented and then briefly compared, when possible, in terms of data rate, determinism, reliability and Quality of Service. Each protocol turns out to have advantages and weaknesses: this implies that the choice of the communication standard to be adopted in a generic future mission will be driven by the specific constraints of the mission itself. Nevertheless, it is possible to get some indications: protocols like SpaceWire and WizardLink, which are currently state of the art and already largely used in a space mission, are not able to fulfil the requirements of future missions. On the contrary, TTEthernet, RapidIO and SpaceFibre are the upcoming protocols that will be able in the near future to provide the required features to next-generation space missions. Within these three, the choice will be made among the various advantages: TTEthernet is built upon solid and well-know protocols with deterministic features, and this can be a key point in terms of reliability and development time; RapidIO can reach a very high data rate and has a valid background in terms of protocol usage; and finally, SpaceFibre is also able to reach very high data rate, is very flexible in terms of Quality of Service configuration and guarantees interoperability with SpaceWire-based devices; moreover, it is an open protocol designed to be fully implemented in hardware. Focusing on multi-laning, SpaceFibre has the advantage to support the grateful degradation of the link. This work deepened the analysis on SpaceFibre, also through the development of various devices and apparatus which enabled the creation of a complete SpaceFibre ecosystem. Although it is not possible to definitively assess whether one of the candidate protocols for future high-speed OBDH is better than the others, we identified several key points that make SpaceFibre particularly appealing for developers of a certain subset of satellite missions (e.g. Earth observation). The rest of the book is intended to be a reference for building and assessing the performance of a generic SpaceFibre-based system).

P. Nannipieri et al., *Next-Generation High-Speed Satellite Interconnect*,
https://doi.org/10.1007/978-3-030-77044-0_7

In Chap. 2, we supported the reader in the understanding of the SpaceFibre protocol in all its different layers; even though the standard contains all the necessary information to build a SpaceFibre-based system, it can be difficult to be understood at first.

In Chap. 3, the various building blocks of a complete SpaceFibre network are illustrated in their architecture and design approach. We first introduced a general architecture of a SpaceFibre CoDec, and a bus functional model of a SpaceFibre communication bus, which can be used as a golden reference in the approach to the verification of the aforementioned CoDec. Then, the design of a SpaceFibre routing switch is presented, also here focusing on the main architectural challenges and design choices where the system designer shall play a major role. Finally, we also illustrated the architecture of a complex Electrical Ground Segment Equipment, which builds on top of the already presented hardware a complex system for the verification and operation of a SpaceFibre-based network. Finally, the SpaceFibre CoDec IP cores available on the market are presented, showing their implementation results for various relevant FPGA technologies, both rad-hardened (RTAX2000, RTG4, Virtex 5) and commercial (ZYNQ 7000).

In Chap. 4, interoperability of independently designed SpaceFibre IP core is tested: to assess the maturity of the SpaceFibre technology. Three independent single-lane SpaceFibre implementations have been intensively tested within one another. A detailed test plan has been produced, aiming at stimulating the core features of the SpaceFibre protocol. The tested devices, from IngeniArs, STARDundee and Cobham Gaisler, demonstrated to be successfully interoperable. The test campaign also offered the chance to compare from the user point of view different link analysers, their features and their user interfaces. The overall indication is that the hardware available on the market is mature enough to be adopted in real scenarios.

In Chap. 5, attention has been paid to SpaceFibre networks. The controlling idea has been to follow the approach that a system engineer will have in the design of a complex SpaceFibre network, providing the necessary tools and references. The first step is to perform a high-level simulation of the network to understand if it has been designed correctly; this can be done thanks to the proposed network simulator (SHINe). Then, we illustrated the design and assembly process of a representative SpaceFibre network with the building blocks presented in Chap. 3. Finally, we introduced a functional and performance test plan to evaluate that network, applicable to any generic network; lastly, the results of the performance and functional analysis are presented, showing results that can be of particular interest to engineers approaching the SpaceFibre world, which will be able to understand the achievable performances of the standard within a network in terms of timing. In particular, jitter results for broadcast messages are particularly encouraging since the jitter itself can be reduced to a single clock cycle with architectural configuration. Such results may be particularly useful in the development of a time-critical system, where a very small jitter may be tolerated. For what concerns packets' latency, it is comparable with older protocol in the case of very small packets: even if the protocol itself has far more bandwidth, the routing switch time for establishing a link between two ports is not negligible. However, for larger packets, we showed that the network can transfer data much more efficiently (exploiting all the available link bandwidth) than in a SpaceWire network, as it should be.

In Chap. 6, we presented a survey on all the existing SpaceFibre-based products. Firstly, we gave an overview of the currently available FPGA technology in the space domain able to host a SpaceFibre interface. Then, we presented and compared the available SpaceFibre CoDecs, routing switches and Electrical Ground Segment Equipment. A comparison with state-of-the-art solutions is presented and an indication of power consumption is given when available. We aimed to provide a valid survey and analysis of the technology available in the literature and on the market. This will help to demonstrate the increasing interest that the SpaceFibre protocol is gaining and, on the other side, we will foster the adoption of the technology.

In conclusion, we assessed the performances and possibilities of the SpaceFibre protocol, demonstrating its performances and its maturity and providing a valid guideline for future system developers. However, it would not be fair to claim that the SpaceFibre protocol is undoubtedly the best option in the foreseeable future for OBDH, as already indicated in Chap. 1. Other candidate solutions certainly have advantages and disadvantages concerning SpaceFibre, and also economic and geopolitical influences may play a major role in determining which technology will be effectively used in the future. Moreover, spacecraft avionics grow constantly in complexity. It is not hard to imagine that in the future not only one single protocol but also a combination of them will be used to combine in a unique network several instruments and building blocks, which are often manufactured independently with very different requirements. We shall also consider that higher level protocols, running on top of the SpaceFibre standard stack, are currently under development. This is possible thanks to the fact that the SpaceFibre standard is published and demonstrated a solid technology baseline for higher level features, which may be required by future spacecraft. ESA is coordinating a working group for the creation of a transaction layer, able to provide more advanced services at the network level, such as

- Time synchronisation messages across the network;
- Standard configuration space;
- Quality of Service and Fault Detection, Isolation and Recovery at the network level.

It would be unwise to write in detail about this topic as all the possible strategies are currently under investigation by the working group. What is certain is that the need for high-level features also in the SpaceFibre network stack will get a standardised answer in the upcoming years. It is easy to think that the first spacecraft employing a SpaceFibre interface will adopt it in its easiest configuration, which is a point-to-point single-lane link. Although this will be a major step forward concerning the currently available solution such as WizardLink, this will not probably be the key aspect fostering SpaceFibre adoption. A possible vision for the future, supported also by the recent SpaceVPX VITA78.1 SpaceFibre standardisation, is to have a system similar to the one shown in Fig. 7.1. In such a scenario, SpaceFibre can be used as a backplane interconnection. By doing so, we can exploit the very high data rate that a Multi-Lane SpaceFibre link can fulfil (up to 100 Gbps for 16 lanes), together with the ease of use of the protocol itself. The SpaceFibre virtual channels can then be used to virtually separate different classes of traffic, coming from the different nodes of the avionics data-handling network. Being SpaceFibre a low-level standard, it will be possible to run a more advanced protocol on top of it (e.g. Ethernet-based protocol). This is just the lift-off of the SpaceFibre adventure. We really hope that this book will help current and future SpaceFibre technology adopters to understand the potential of this communication protocol.

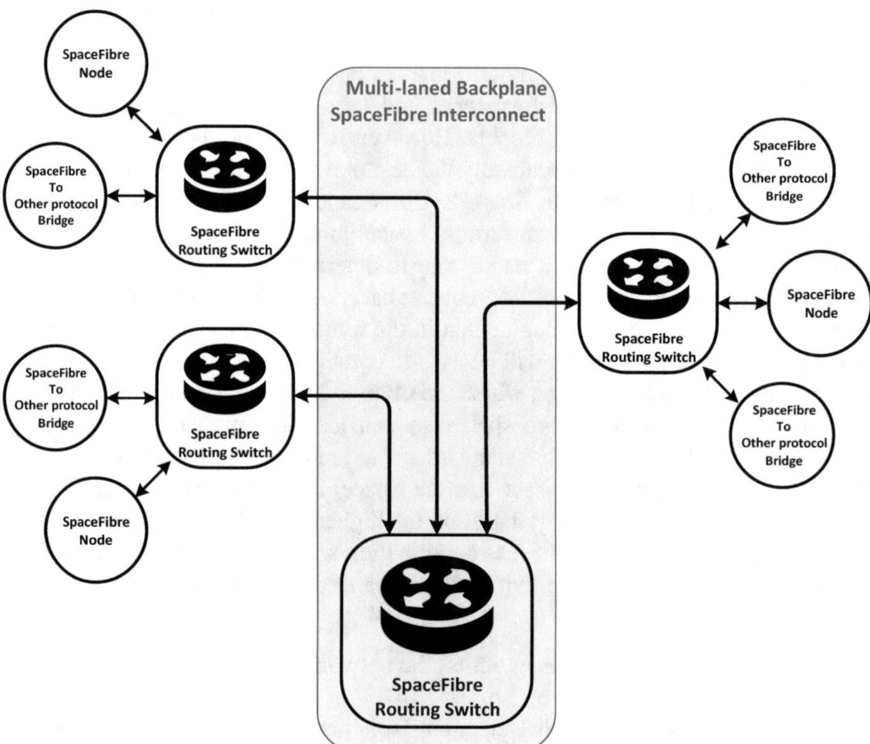

Fig. 7.1 Potential future avionics system based on a SpaceFibre Multi-Lane link as backplane interconnect

Appendix A
Reduced SpaceFibre CoDec: Design and Technology Mapping

This chapter proposes a modified implementation of the SpFi standard, R-SpaceFibre (R-SpFi), which has been designed to reduce hardware resources while keeping high data rate capability and flow control, as presented in [7]. Firstly, attention is given to the motivations that can lead to adopting a modified version of the SpFi of the standard. Then, details about the architecture modifications are explained focusing on the trade-offs between Data-Link layer complexity and the reduction of the standard features. Implementation results showing a reduction of more than 40% of logical resources required per single interface are presented. Finally, a CubeSat use case scenario is presented.

A.1 The Need of a Reduced Version of the SpaceFibre Standard

In particular cases, also SpFi may be over-specified. For instance, the data retry feature is not even removable in payloads characterised by streaming of a large amount of data that do not contain critical information (e.g. high-resolution imaging payloads, where it is acceptable to lose data frames in case of upsets). Moreover, for small satellites and CubeSats [1], where the size of components particularly matters, both high-speed interface and data processing unit are to be embedded in the same FPGA.

R-SpaceFibre offers the possibility to exploit SpaceFibre high-speed capability and use optical fibre as a physical link. It provides features that are comparable with WizardLink technology, but with the advantage to be based on an open protocol, also guaranteeing compatibility with SpaceFibre and SpaceWire.

An attempt to design a SpFi endpoint with reduced complexity has already been presented in [9], where an architecture to reduce area occupation has been proposed. Unfortunately, the lack of details on the hardware configuration (e.g.

P. Nannipieri et al., *Next-Generation High-Speed Satellite Interconnect*,
https://doi.org/10.1007/978-3-030-77044-0

target frequency, implementation of data scrambling, VCs size, etc.) combined with the fact that the standard significantly developed since 2014 make it inappropriate to compare analytically our proposed implementation with the one contained in [9]. The work presented in this appendix aims at proposing a novel reduced architecture of the SpFi standard, R-SpFi, that optimises resource utilisation at the expense of reduced features. Furthermore, the proposed modifications to the standard are intended to be compatible with a fully compliant implementation of it. R-SpFi has been realised starting from the IngeniArs SpFi CoDec IP. Since the work presented in this appendix aims at presenting a modified and reduced version of the SpFi standard, the results presented are all referred to as the same full SpFi CoDec (from IngeniArs). The obtained resource reduction is then theoretically achievable also on other different implementations.

A.1.1 SpaceFibre Data-Link Layer

The Data-Link layer of a fully compliant version of the protocol is described in Sect. 2.5 and resumed here for comparing it with the R-SpaceFibre alternative solution. Figure A.1 shows a simplified architecture block of the SpFi Data-Link

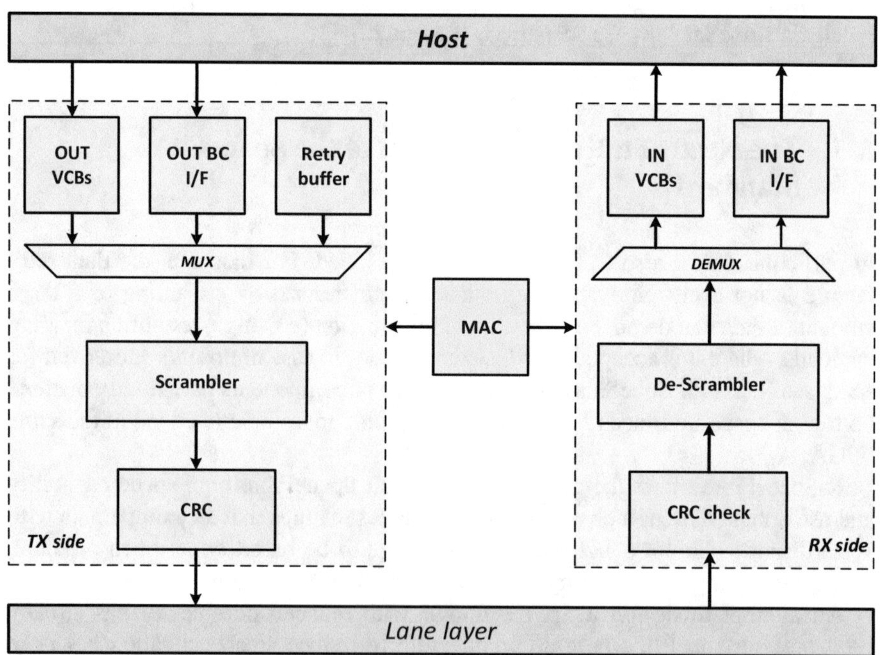

Fig. A.1 SpFi Data-Link layer block diagram [7]

layer. It is composed of a *MAC*, a transmission block (*Tx Block*) and a receiving block (*Rx Block*).

The data interface with the host is realised through VCs, both on the RX side (*OUT VCs*) and on the TX side (*IN VCs*). They allow handling independent flows of information across a single physical link. There are up to 32 different VCs available. A special interface handles the transmission (*OUT BC I/F*) and reception (*IN BC I/F*) of BC messages.

The *MAC* is the core of the Data-Link layer: it manages data framing and it is responsible for synchronising the operations between transmitting and receiving sides. Moreover, the *MAC* implements the *QoSLabel*: it continuously schedules which VC shall send data based on different parameters such as priority and reserved bandwidth.

The Data-Link layer also implements the Retry Buffer block, which re-sends data in case the far-end of the link detects a communication error and asks for retransmission.

Data or BCs are read out from the selected VC or the corresponding Retry Buffer and framed appropriately. Then, the data frame could be scrambled to reduce electromagnetic emissions and eventually, the CRC [5] block computes a numeric field to be added at the end of each data frame, broadcast message and control words that require it. The far-end will recompute independently this value and will compare it with the received one, to detect accidental errors that occurred during the communication process and promptly ask for retransmission of corrupted frames or control words.

Vice versa, on the Rx side, received data frames, BC messages and control words are handled. First, the CRC Check block controls the correctness of the CRC value embedded in the received data. In case of errors, a retry request is sent to the far-end. If no errors are detected, the data frame may be de-scrambled and stored in the appropriate *IN VCs*, whereas broadcast messages are stored in the *IN BC*. Control words are consumed for managing the communication process. For detailed information, please refer to the SpFi standard [4].

A.1.2 R-SpaceFibre Data-Link Layer

This section describes the proposed architecture of the R-SpFi Data-Link layer. A simplified architectural block diagram of the layer is shown in Fig. A.2. R-SpFi major innovation is the exclusion of the retry mechanism from the protocol stack. The first consequence is that neither the Data Retry Buffer nor the Broadcast Retry Buffer nor the FCT Retry Buffer has to be included in the R-SpFi Data-Link layer, with consistent savings of logical resources. This comes at the price of no data retry if an error occurs with consequent data loss. However, the system is designed to continue the communication process, and a corrupted datum is passed to a higher level of the protocol, but any lost FCT must be avoided or leads to link initialisation. Target applications such as SAR imagers can withstand the corruption of a single

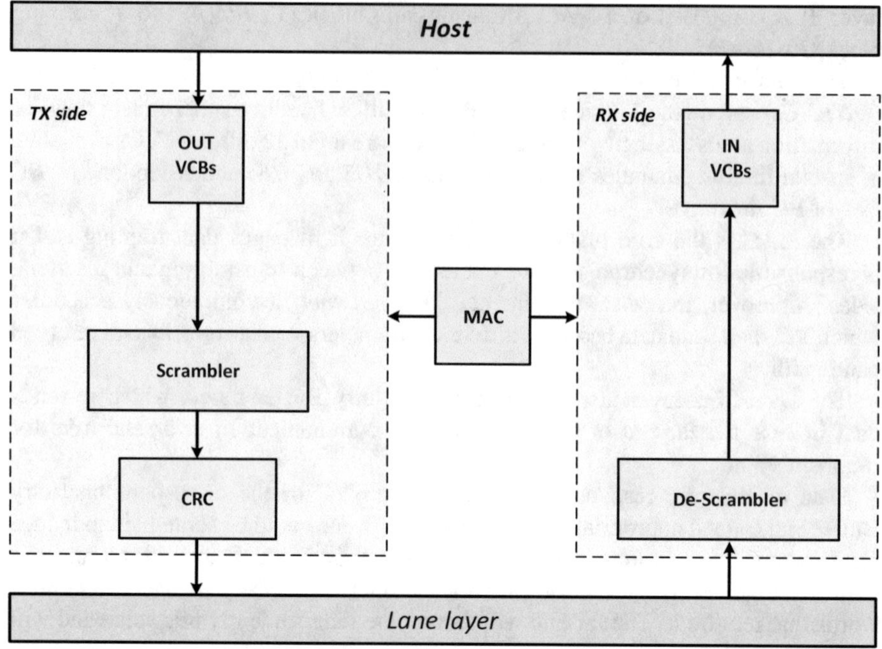

Fig. A.2 R-SpFi Data-Link layer block diagram [7]

bit or the loss of a data word. The SpFi CoDec properly works with a maximum BER of 10^{-5} [6]. However, typical BERs for satellite applications are in the order of 10^{-10}, which is much lower than the worst case handled by the SpFi standard [4]. The application system engineer will decide whereas complexity reduction at the price of data retry removal is acceptable or not for its application, depending on link criticality, orbit BER, etc.

R-SpFi is designed mainly for point-to-point communication. For this reason, the broadcast service, responsible for sending short messages to all the nodes of a network, is unnecessary. Thus, *In* and *OUT BCs* have been excluded from the architecture. The removal of retry and broadcast services implies that also all QoS mechanisms (e.g. VCs scheduling and bandwidth reservation) become less complex, and thus the MAC can be simplified.

The CRC Check block has been removed from the Rx Block: consequently, the IP core is no more able to signal to the far-end if received data frames or control words are corrupted. The near-end needs to acknowledge also corrupted data frames so that compatibility with a full implementation of the standard at the far-end is maintained. More details about R-SpFi compatibility with a full SpFi interface are provided in Sect. 2.3. On the contrary, the CRC on the Tx Block cannot be removed: it has to be maintained to be compatible with the SpFi protocol. Indeed, a standard SpFi endpoint always performs CRC Check on received data frames and control words; without it, the CoDec would automatically consider all received data as corrupted.

Nevertheless, the CRC block could be excluded from the Tx Block if the R-SpFi just needs to communicate with R-SpFi nodes. The proposed design includes optional data scrambling on both Tx and Rx sides.

The SpFi standard establishes that the number of VCs shall be between 1 and 32. R-SpFi can handle an arbitrary number of VCs, but to reduce the impact on resource usage, the number of VCs shall be limited to 1 or 2. In particular, by choosing to use only 1 VC, the QoS is drastically reduced, together with the multiplexing logic between VCs and the lower part of the Data-Link layer. Both standard and literature [9] suggest that the QoS could be eliminated in implementation of just of 1 or 2 VCs, but this implies that babbling node protection is not provided [4].

Finally, the VC dimension is set to be 256 data words, which is the minimum dimension according to the standard. In conclusion, the most relevant design changes in R-SpFi development are the removal of

- Retry Buffers, from the Tx Block;
- Broadcast service, from the entire Data-Link layer;
- CRC Check block, from the Rx Block;
- Several QoS mechanisms, from the MAC.

A.1.3 Fault Tolerance and Compatibility with Full SpFi Interfaces

R-SpFi is meant to be used as a lightweight high-performance high-speed link. It is a slightly modified version of the SpFi protocol, fully interoperable with a SpFi endpoint but compliant just to a subset of the SpFi standard requirements, to reduce system complexity. When two R-SpFi endpoints communicate together, no restrictions on protocol operations arise, also in the case of SEUs bit-flip errors happening on the link. The received corrupted packet will be handled at higher levels of the protocol, application-dependent. However, it is possible to connect R-SpFi and SpFi endpoints. Coherent data transmission and reception are guaranteed, discarding received broadcast messages and ignoring retry requests. R-SpFi is programmed to discard any retry request from a far-end. Consequently, if an error occurs during the transmission of a packet from a full SpFi interface to a reduced SpFi endpoint, no retry request is given and the received corrupted packet is transmitted at the higher level of the protocol, where it will be handled according to the application requirements. Indeed, R-SpFi has been designed for communication links involving payloads streaming a large amount of data that do not contain safety-critical information. On the other hand, when a SEU occurs in the transmission of a packet from an R-SpFi interface to a full SpFi interface, the second one will constantly try to ask for data retransmission, blocking the traffic on the SpFi link. This can be easily handled by automatically asserting a reset signal to restore the communication flow.

A.2 Hardware Implementation

An R-SpFi implementation has been tested and validated on a Xilinx ZC706 evaluation board. The SoC mounted embeds one Zynq-7000 XC7Z045-2FFG900C FPGA, one Cortex A9 processor and up to 16 GTX transceivers. The hardware set-up is shown in Fig. A.3: the programmable logic of the FPGA is used to implement two SpFi CoDecs, which communicate using 2 GTXs. The transceivers are externally connected through a Xilinx FMC XM104 expansion board, using e-SATA cables. The Cortex A9 generates and consumes packets to be transmitted or received over the SpFi link, as described in [3]. The system automatically checks whether all received packets are correct.

In order to validate R-SpFi, two configurations have been tested:

- CoDec 0 is an R-SpFi CODEC and CODEC 1 is a SpFi CODEC (CONFIG1).
- CoDec 0 and CoDec 1 both are R-SpFi CoDecs (CONFIG2).

The system has been intensively stimulated through an appropriate hardware validation plan, focusing on Data-Link layer functionalities. In particular, QoS features, VCs scheduling, and configurations with and without data scrambling have been tested. Furthermore, errors (i.e. bit-flips) have been injected into the data stream to verify the behaviour of the CoDecs both in CONFIG1 and in CONFIG2.

In CONFIG1 tests, a deadlock arises when the SpFi node sends too many retry requests to the R-SpFi node, as described in Sect. 2.2. In particular, tests show that the time between two consecutive deadlocks increases decreasing packet

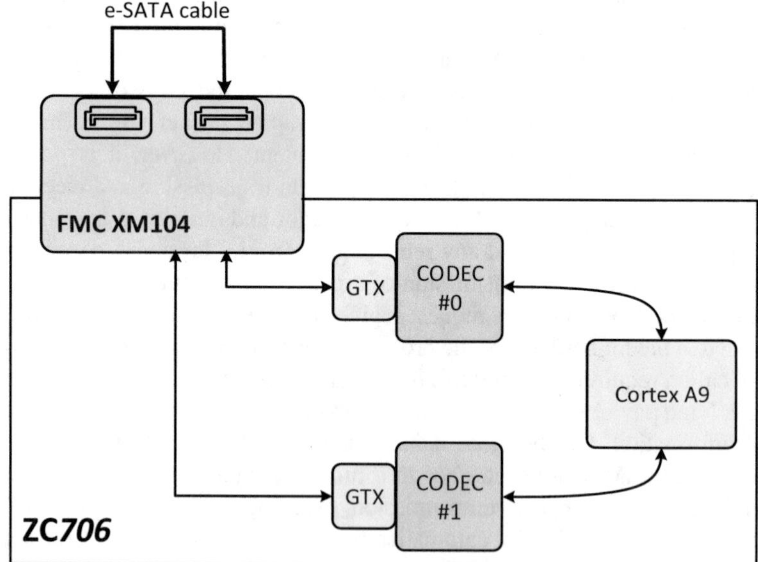

Fig. A.3 Architectural diagram of the test bed [7]

dimension. The FCT counter is incremented by a number equal to the number of words composing a correctly received packet. CONFIG2 tests show data packet loss as expected, but no deadlock situation arises because R-SpFi considers all received packets valid. The performed tests validated R-SpFi: the reduced CoDec demonstrated to be able to communicate with a version of the CoDec complaint with the SpFi standard, as well as with another reduced implementation.

A.2.1 Resources Utilisation and Power Consumption

In this section, resource utilisation after carrying out the logic synthesis, implementation and place and route on different space-grade FPGAs are presented. In particular, the number of LUTs, FFs, and BRAMs necessary to implement one R-SpFi CoDec is reported.

Table A.1 shows the resource utilisation for the R-SpFi CoDec on a Xilinx Zynq-7000, a Xilinx Virtex 5, a Microsemi RTG4 and a Microsemi RTAX2000. These numbers refer to an R-SpFi implementation with a single 256-word VC and 62.5 MHz clock frequency, achieving a data rate of 2.5 Gbps (SpFi, and consequently R-SpFi, shall transmit one 40-bit word per clock cycle).

Finally, Table A.2 shows a comparison between the hardware resources of R-SpFi and IngeniArs SpFi CoDec on a Microsemi RTAX2000 and a Microsemi RTG4. The CoDecs implemented still has a single 256-word VC, and the target frequency is 62.5 MHz for RTG4 and only 40 MHz for the RTAX2000, due to the lower performances of the FPGA, with a maximum data rate of 1.6 GHz. This is acceptable as the aim of the work here described is to demonstrate the possibility to reduce hardware resources needed by a SpFi end-node. It is clearly observable that R-SpFi considerably reduces (more than 40%) the hardware resources needed for its implementation on both RTAX2000 and RTG4. R-SpFi implementation allows to save consistent part of synthesizable logic; this is particularly important for FPGAs like the RTAX2000 which have a limited number of hardware resources respect to bigger size FPGAs like with RTG4.

Table A.1 R-SpFi CoDec resource utilisation on different FPGAs

FPGA	VC	LUT	Util% LUT	FF	Util% FF	BlockRAM-FIFO
Zynq-7000	1	1675	0.73%	1791	0.4%	1.5
	2	2248	1%	2345	0.5%	2
Virtex 5	1	1682	2.05%	1103	1.35%	2
	2	2662	3.25%	1653	2.02%	4
Microsemi RTG4	1	2561	2.07%	1227	1.19%	16
	2	4122	3.47%	1892	2.00%	32
Microsemi RTAX2000	1	3479	16.18 %	1530	14.23%	4
	2	5356	24.91%	2641	24.56%	8

Table A.2 SpFi vs R-SpFi resources utilisation on Microsemi RTAX2000 and RTG4

	RTAX2000		RTG4		Zynq-7000		Virtex 5	
	LUT	Util%	LUT	Util%	LUT	Util%	LUT	Util%
SpFi	5830	27.11%	4632	3,98%	2800	1.28%	2919	3.56%
R-SpFi	3479	16.18%	2561	2,07%	1675	0.73%	1682	2.05%
% Reduction	2351	40.33%	2071	44,71%	1125	40.18%	1237	42.39%
	FF	Util%	FF	Util%	FF	Util%	FF	Util%
SpFi	2605	24.23%	1994	2,24%	2441	0.56%	1702	2.08%
R-SpFi	1530	14.23%	1227	1,19%	1791	0.4%	1103	1.35%
% Reduction	1075	41.27%	767	38,46%	650	26.63%	599	35.20%
	BRAM	Util%	BRAM	Util%	BRAM	Util%	BRAM	Util%
SpFi	12	18.75%	39	1,10%	5.5	1.28%	7	2.35%
R-SpFi	4	6.25%	16	0,45%	1.5	0.28%	2	0.67%
% Reduction	8	66.66%	23	58,97%	4	72.72%	5	71.42%

Table A.3 R-SpFi and SpFi power consumption

	Dynamic power [mW]	Static power [mW]	Total power [mW]
R-SpFi	49	199	248
SpFi	55	199	254

Post-layout simulations have been carried out on a Xilinx Zynq-7000 to estimate the power consumption of both R-SpFi and SpFi IP cores. The results have been obtained with 200 μs simulations, corresponding to the transmission of 520 kBit of data (divided into 36-bit data words). The CoDecs have been implemented with 1 VC and the operating frequency is set at 62.5 MHz. Results are shown in Table A.3. Power consumption results similar to the two implementations. R-SpFi does not implement the broadcast service and retry mechanism. However, in a standard SpFi communication link, those mechanisms are rarely activated. Thus, the small difference between the two power consumption can be attributed to the fact that SpFi CoDec computes the CRC code on received data frames.

Appendix B
A WizardLink Equivalent Interface

This appendix chapter presents the work done to use the already designed lower Lane layer building blocks of the SpaceFibre interface to design a WizardLink equivalent FPGA IP [2]. The current mission with stringent data rate requirements, which SpaceWire is not able to fulfil no more and which do not already rely on SpaceFibre solutions for its lack of heritage, is currently leading to a partial adoption by the industry and by space agencies of a non-standardised solution, the TLK2711 ASIC chip from Texas Instrument, which implements lower communication protocol levels of the Open System Interconnection stack (WizardLink), leaving to the user the implementation of higher level stack layers. Consequently, in the space community, the need for hardware to test these systems has grown. In the following, we present an implementation of a WizardLink TLK2711 equivalent circuit on a commercial FPGA, mainly made of the lower Lane layer building blocks of a SpaceFibre interface: such a design realised on HDL portable code can boost the development of TLK2711 compatible ground test equipment on commercial devices. Moreover, it enables to design EGSEs with configurable interfaces, where the higher level SpaceFibre protocol stack can be activated/deactivated depending on the user needs.

B.1 The Need and Potential of a WizardLink Equivalent FPGA IP

Recent trends and evolution in space missions push forward the request of higher on-board communication data rates [11]. Currently, on-board data-handling high data rates (in the order of the Gbps) are obtained only with not-standardised interfaces. However, the use of a SerDes is must to build either completely custom protocols, parallel LVDS or the widely used WizardLink (up to 2 Gbps). WizardLink protocol is based on a chipset solution from Texas Instrument called TLK2711 [10]

P. Nannipieri et al., *Next-Generation High-Speed Satellite Interconnect*,
https://doi.org/10.1007/978-3-030-77044-0

(see Fig. B.1), providing bi-directional point-to-point communication, 16-bit data bus, 8B/10B encoding and parallel-to-serial/serial-to-parallel registers. The achievable data rate is up to 2Gbps.

In the years between 2010 and 2020, many space missions adopted the WizardLink high-speed interface, such as Sentinel-1, COSMO-SkyMed Second Generation and ALOS-2. Also, many space missions currently in the design and assembly phase rely on the WizardLink protocol. As with any other on-board communication interface, the TLK2711 is a vital part of EGSE of WizardLink-based spacecraft. At the present moment, without any TLK equivalent device to be embedded in an FPGA, in case a system engineer needs to test a WizardLink-based system, it would require a rather complex set-up, as shown in Fig. B.2.

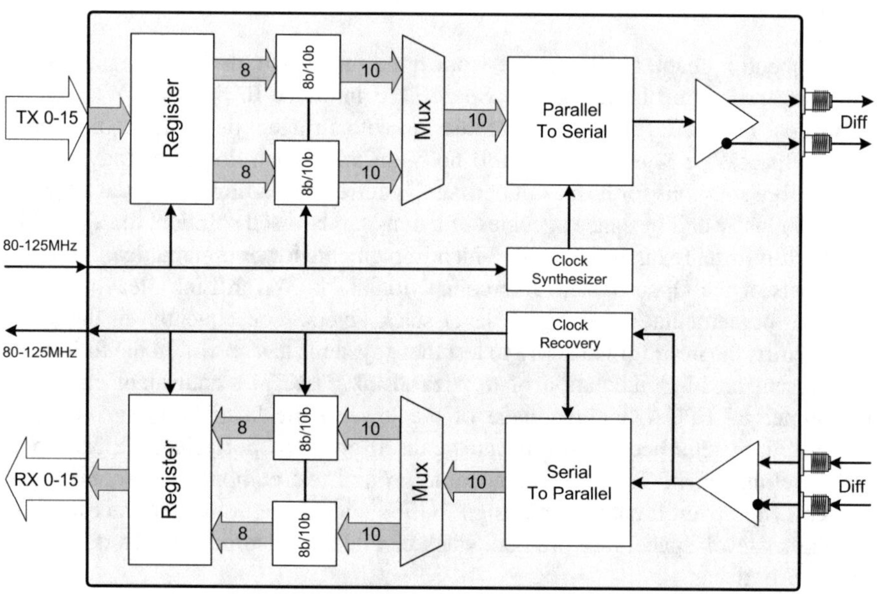

Fig. B.1 TLK2711 high-level block diagram [2]

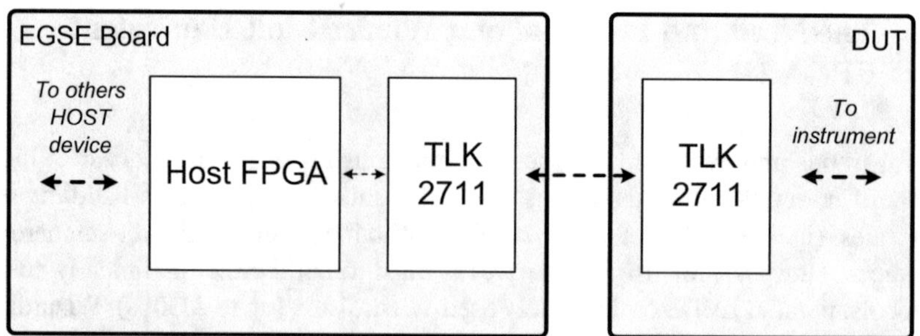

Fig. B.2 Classical test set-up for DUT with WizardLink interface

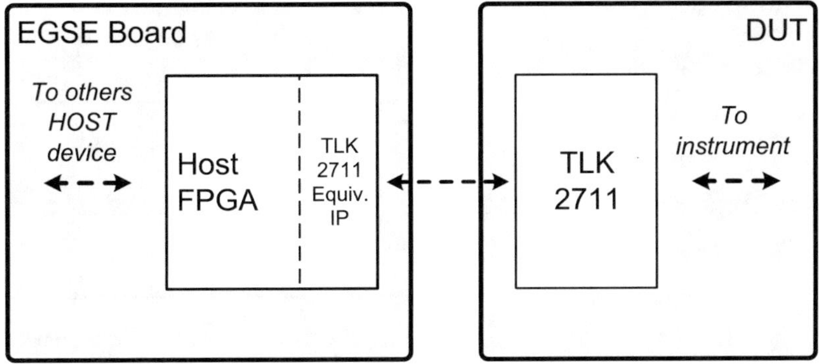

Fig. B.3 Test set-up with proposed architecture for DUT with WizardLink interface

With this configuration, three different devices are used: the EGSE requires a TLK2711 IC itself as an interface, to be connected with an FPGA that will do the data processing. This results in higher costs and complexity: more components are used and a more complex custom printed circuit board needs to be produced. On top of that, at the time of writing, Texas Instruments are producing only the space-qualified version (SP) of the TLK2711; therefore, a radiation-hardened chip should be included in an EGSE without any radiation requirements. All these factors contribute to lower the competitiveness of such systems on the market. If we introduce a TLK2711 equivalent IP that can be embedded in common FPGAs with SerDes, it is possible to realise the same test set-up as shown in Fig. B.3. The DUT TLK2711 interface is connected with the high-speed serial interface of the FPGA, which embeds both the TLK2711 equivalent RTL IP and the rest of the EGSE system.

The core of the proposed system is the TLK2711 equivalent IP. It exactly emulates the behaviour of a WizardLink node on a technology-independent solution, portable on different FPGA platforms. With these capabilities, a potential system integrator will be able to directly interface the programmable logic section of the EGSE, provided that it embeds high-speed serial link capabilities, directly to a WizardLink interfaced DUT.

B.2 Design and Validation of WizardLink Equivalent IP

To achieve full compatibility with a WizardLink interface, the TRK2711 equivalent RTL IP, we shall propose the following:

- Perform 8B/10B encoding to 20-bit word on the TX side;
- Perform 8B/10B decoding to 16-bit word on the RX side;
- Perform comma alignment of symbols parallelised by the SerDes;

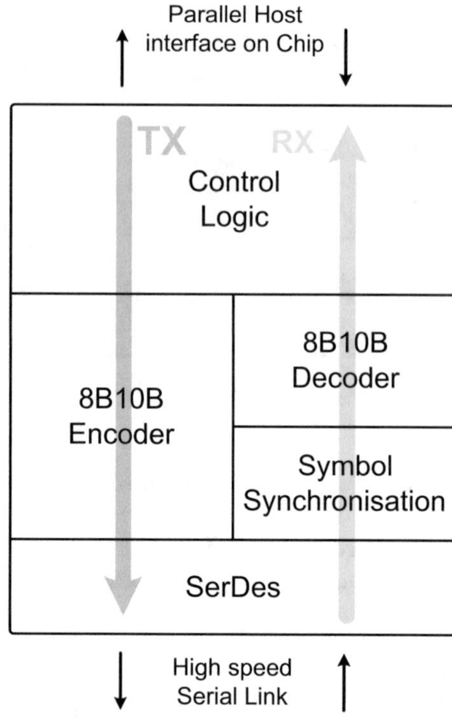

Fig. B.4 TLK equivalent IP block scheme

- Serialise 20-bit word on TX side and de-serialise 20-bit word on the RX side;
- Properly signals error situations to the host, e.g. output K31.7 symbol on both bytes in case of loss of signal.

The IngeniArs design, presented in this chapter, maps the features described in Fig. B.1 on commonly available FPGA resources. The proposed block scheme is shown in Fig. B.4.

The system is composed of the following blocks:

- A SerDes Block, responsible for shifting parallel data paths into serial high-speed links, both in transmission and in reception. Since the clock is not transmitted, the SerDes shall recover the RX clk signal from the input serial data stream.
- On RX Side, a block called *Symbol Synchronisation* is responsible for properly aligning the 20-bit words outputted by the de-serialiser, using unique bit sequences called commas.
- An 8B/10B encoder/decoder. The advantages of this form of encoding are redundancy on symbols which make easy to understand whether an error occurred and easy clock recover due to the high number of transition between zeroes and ones.

- A compact control block is built on top of the other section of the IP. It ensures that all the high-level features of the TLK2711 are properly emulated, e.g. the way errors are handled.

It is worthy to observe that all these features are carefully described in Sect. 2.3, within the lower Lane layer of SpaceFibre. WizardLink is a subset of the SpaceFibre protocol stack, precisely a subset of its Lane layer, with little addition of small control logic. It is therefore strategical to start working with it to support the current spacecraft mission and then expand the protocol stack to support all the SpaceFibre functionalities. To validate the proposed design, we implemented it on an FPGA board. Since the first need for such a system is focused on EGSEs, we targeted the National Instruments PXIe 6591 high-speed serial instrument board. The main reason and stricter requirement fulfilled is the availability of GTX transceivers, which are integrated into the Xilinx Kintex 7 chip-on-board. Moreover, it is a common reference platform in EGSE architectures, equipped with Xilinx devices, therefore easily portable. PXIe 6591 GTX transceivers are capable of sustaining serial links with throughput up to 12.5 Gbps, largely exceeding the 2 Gbps requirement. To implement the system design into an NI PXIe FPGA board, it is necessary to exploit the use of the Socketed component-level IP (Sck Clip) asset in LabVIEW FPGA workflow [8].

B.3 Results

The TLK2711 Equivalent system has been designed with the Xilinx Vivado design tool, according to the FPGA chip-on-board of the target PXIe module. Also in Vivado, system design has been synthesised for the very same target FPGA chip, and then product results have been exported as EDIF files. Netlists have been imported into the LabVIEW project as sources for the Sck Clip module.

The Sck CLIP methodology has been intentionally employed not to import a small custom HDL design in a larger LabVIEW FPGA project, but to wrap a whole system design under the LabVIEW environment instead. In Table B.1, the results in terms of occupied resources are presented.

Figure B.5 provides an overview of the TLK2711 equivalent system. The NI PXIe 6591 Peripheral Module is hosted into a PXIe compatible chassis, along with the user PXI controller. The on-board FPGA chip can communicate through the PXIe bus thanks to a dedicated interface. Still, front panel connectors can be accessed thanks to the Sck Clip methodology. The described design flow, necessary to integrate the IP core within the PXI platform, and the obtained resource

Table B.1 TLK2711 equivalent resource utilisation on Xilinx Kintex 7

LUT	REG	GTX primitive	MMCME2 or PLL
476	596	1	2

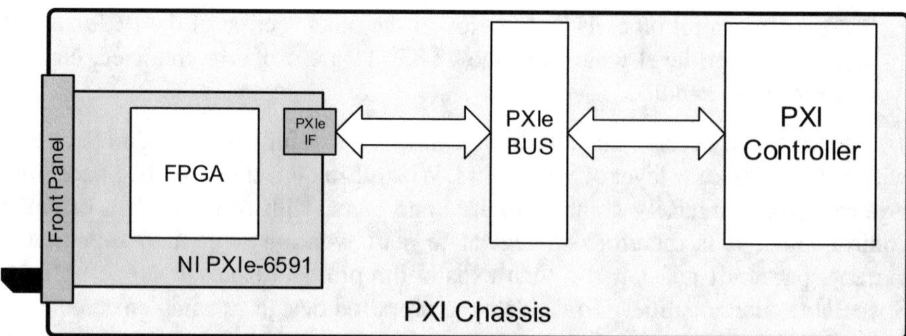

Fig. B.5 Block diagram of the proposed system

utilisation, highlights once more the low number of resources needed to implement such a system, which leaves plenty of space on the FPGA board to build specific EGSE functions.

References

1. Cubesat Design Specification Rev. 12. (2009). Available online: http://www.cubesat.org/resources/.
2. Dello Sterpaio, L., Marino, A., Nannipieri, P., & Fanucci, L. (2018). A PXI based implementation of a tlk2711 equivalent interface. In: *International Conference on Applications in Electronics Pervading Industry, Environment and Society*, pp. 407–413. Springer.
3. Dinelli, G., Davalle, D., Nannipieri, P., & Fanucci, L. (2018). A SpaceFibre multi lane codec system on a chip: Enabling technology for low cost satellite EGSE. In: *Proceedings of the PRIME 2018 14th Conference on Ph.D. Research in Microelectronics and Electronics*, Prague, Czech Republic, 2–5 July 2018, pp. 173–176.
4. European Cooperation for Space Standardisation. (2019). *SpaceFibre – Very high-speed serial link, ECSS-E-ST-50-11C*. European Cooperation for Space Standardisation.
5. Koopman, P., & Chakravarty, T. (2004). Cyclic redundancy code (CRC) polynomial selection for embedded networks. In: *International Conference on Dependable Systems and Networks*, pp. 145–154.
6. Leoni, A., Dello Sterpaio, L., Davalle, D., & Fanucci, L. (2016). Design and implementation of test equipment for SpaceFibre links: SpaceFibre, short paper. In: *2016 International SpaceWire Conference (SpaceWire)*, Yokohama, Japan.
7. Nannipieri, P., Davalle, D., Dinelli, G., & Fanucci, L. (2019). Design of a reduced SpaceFibre interface: An enabling technology for low-cost spacecraft high-speed data-handling. *Aerospace, 6*(9), 101.
8. Nannipieri, P., Marino, A., Dello Sterpaio, L., & Fanucci, L. (2019). Exploiting LabViewFpga socketed clip to design and implement soft-core based complex digital architectures on PXI FPGA target boards. In: *2019 16th International Conference on Synthesis, Modeling, Analysis and Simulation Methods and Applications to Circuit Design (SMACD)*, pp. 193–196. IEEE.

9. Rowlings, M., & Suess, M. (2014). An experimental evaluation of SpaceFibre resource requirements. In: *2014 International SpaceWire Conference (SpaceWire)*
10. Texas Instruments. (2001). Tlk2711 1.6 to 2.7 gbps transceiver datasheet.
11. Thompson, P. T., Corazza, G. E., Vanelli-Coralli, A., Evans, B. G., & Candreva, E. A. (2011). 1945–2010: 65 years of satellite history from early visions to latest missions. *Proceedings of the IEEE, 99*(11), 1840–1857.

Glossary

Fault Detection Isolation and Recovery is the system responsible for identifying accidental errors occurred during the communication process and for the retransmission of the corrupted data.

Space AVionics Open Interface aRchitecture (SAVOIR) is an initiative to federate the space avionics community and to work together in order to improve the way that the European space community builds avionics sub-systems.

European Cooperation for Space Standardisation is an initiative established to develop a set of standards to be used in all European space activities.

Cyclic Redundancy Check is an error-detecting code used for identifying accidental error occurred during the communication process.

Scrambling is a technique used for reducing the electromagnetic emissions produced by the transmission of data over a physical medium.

SpaceFibre is a high-speed link technology developed to be part of the future on-board data-handling sub-systems.

Simulator for High-speed Network (SHINe) is a discrete event simulator supporting both SpaceFibre and SpaceWire protocols and it is entirely based on the open-source framework OMNeT++.

Data Link layer is the protocol layer principally responsible for SpaceFibre Quality of Service and Fault Detection Isolation and Recovery system.

Lane layer is the protocol layer responsible for establishing the communication between the two ends of a SpaceFibre link.

Multi-Lane layer is the protocol layer responsible for managing the operation of up to 16 lanes operating in parallel.

Lane is the physical connection between two SpaceFibre ends.

Link is the physical connection between two SpaceFibre ends which includes 1 or more lanes.

Multi-Lane link is a link that comprehends two or more lanes.

Data rate is the rate of transmission of data over a link.

P. Nannipieri et al., *Next-Generation High-Speed Satellite Interconnect*,
https://doi.org/10.1007/978-3-030-77044-0

R-SpaceFibre is the reduced version of the SpaceFibre protocol stack that aims at reducing the hardware resources required for its implementation.

Rad-hard FPGAs are devices specifically designed to operate in the space radiation environment.

SpaceART is a complete testing solution developed by IngeniArs for high-speed links in space applications. SpaceART supports both SpaceWire (SpW) and SpaceFibre (SpFi) standards.

StarFIRE MK3 is a test and development unit that can emulate, stimulate, debug and validate SpaceFibre enabled equipment.

Quad SpaceFibre card is an advanced PCIe interface, supporting SpaceFibre simulation.

Multi-Lane layer recovery time is the time interval necessary to the Multi-Lane layer of a SpaceFibre interface to recover from the disconnection of a lane and start again to exchange data.

Electrical Ground Segment Equipments are tools used by satellite and subsystem manufacturers and integrators to test and validate electrical functions of the satellite on the ground before launch.

Virtual Channel is an independent channel that can carry information across a single link in parallel with other independent information carrying channels.

Broadcast Channel is an independent channel that can carry broadcast messages across a single link.

Flow Control is the process of managing the rate of data transmission between two nodes.

Virtual Network is a logical network that runs in parallel with other logical networks over a single physical network.

Routing switch comprises several SpaceFibre ports and a switch matrix that switches packets arriving on one port out of another port according to the destination address of the packet and the contents of a routing table, which validates and broadcasts broadcast messages out of all of the ports except the one on which the broadcast message arrived and which includes a configuration port for configuring the ports and the routing switch itself. Please note that the term "router" is synonymous.

Physical layer is the protocol layer that specifies the cables, connectors, cable assemblies, line drivers, line receivers, serialisation and de-serialisation.

Network layer is the protocol layer that is responsible for packet forwarding including routing through intermediate routers.

Management layer is the protocol layer that specifies status and control signals of a SpaceFibre device. Please note that the term "management information base" is synonymous.

Quality of Service is the ability to provide different priorities to different applications, users or data flows or to guarantee a certain level of performance to a data flow.

Bus Functional Model is a non-synthesisable software model of an integrated circuit component having one or more external buses.

SpaceWire is a spacecraft communication network based in part on the IEEE 1355 standard of communications.

WizardLink is a family of transceivers produced by Texas Instruments, and in particular, for space applications, the TI TLK2711 is typically referred.

TTEthernet is the Time-Triggered Ethernet (SAE AS6802), a standard defining a fault-tolerant synchronisation strategy for building and maintaining synchronised time in Ethernet networks, and outlines mechanisms required for synchronous time-triggered packet switching for critical integrated applications.

RapidIO is a high-performance packet-switched interconnect technology.

On-board data-handling sub-system of a spacecraft is the subsystem that carries and stores data between the various electronic units and the ground segment, via the telemetry, tracking and command (TT&C) subsystem.

Universal Verification Methodology is a standardised methodology for verifying integrated circuit designs.

Index

P. Nannipieri et al., *Next-Generation High-Speed Satellite Interconnect*,
https://doi.org/10.1007/978-3-030-77044-0

Printed in the United States
by Baker & Taylor Publisher Services